Neopragmatic Horizontal Geographies

Volume 3

Series Editors

Olaf Kühne, Geographisches Institut, Raum H214, Universität Tübingen, Tübingen, Germany

Karsten Berr, Department of Geography, Tuebingen University, Tübingen, Germany

Lara Koegst, Tübingen, Germany

Central aspects of philosophical neopragmatism have been made available for integrative spatial research in recent years. Following classical pragmatism, which measures the quality of cognition by its suitability for problem solving, rather than by philosophical or moral principles, neopragmatism also integrates considerations of language theory. The central element of this meta-theoretical approach is the plurality of perspectives appropriate to the subject matter. This plurality gains particular relevance in the study of complex objects. Especially in the case of spatial syntheses, as they are carried out in 'horizontal geographies' (e.g. in the form of region, city or landscape), a multitheoretical, also multimethodical and inter- and transdisciplinary approach is required—following the neopragmatic approach. Due to the inter- and transdisciplinarity of the approach, there is a special focus on innovative forms of presenting the results of research, following the neopragmatic approach of horizontal geographies.

The aim of the series is first of all to elaborate the meta-theoretical fundamentals of 'neopragmatic horizontal geographies', which form the justifying basis for the triangulation of theories, research methods derived from them, and last but not least methods of representation. Studies that operationalize the meta-theoretical framework with different spatial as well as theoretical and methodological shear points illustrate the potentials of the approach for theory-driven horizontal geographic research.

Olaf Kühne · Irina Silina · Karsten Berr

Sustainable Development and Life Chances

A Redescription From a Neo-Pragmatist Perspective

Springer

Olaf Kühne
Fachrichtung Geographie
Eberhard Karls Universität Tübingen
Tübingen, Baden-Württemberg, Germany

Karsten Berr
Fachrichtung Geographie
Eberhard Karls Universität Tübingen
Tübingen, Baden-Württemberg, Germany

Irina Silina
Institute of Climate and Energy Systems (ICE)
Forschungszentrum Jülich GmbH
Jülich, Nordrhein-Westfalen, Germany

ISSN 2731-9881 ISSN 2731-989X (electronic)
Neopragmatic Horizontal Geographies
ISBN 978-3-032-20181-2 ISBN 978-3-032-20182-9 (eBook)
https://doi.org/10.1007/978-3-032-20182-9

This Springer imprint is published by the registered company Springer Nature Switzerland AG
The registered company address is: Gewerbestrasse 11, 6330 Cham, Switzerland

Contents

Chapter 1
Introduction

Sustainable development has become a key concept in society's self-description. At the same time, tensions are increasingly arising in the context of sustainable development. While sustainable development originally aspired to combine ecological, social, and economic concerns, the concept has become a projection screen for diverging expectations, normative overloads, competing visions of the future, and a medium for social "culture conflicts" in public and political discourse (e.g., Kühne et al., 2024; Lehmann, 2024; Stegemann, 2023). Conflicts over energy transition, mobility, nutrition, and land use illustrate that sustainable development requires more than just technical or economic solutions. Rather, sustainable development is linked to cultural, communicative, and symbolic debates about what can be considered desirable futures, not least in terms of how societies make these futures narratable (among many: Hoeft et al., 2017; Kamlage et al., 2020; Laborgne & Radtke, 2023; Mölders, 2025; Radtke, 2025; Radtke & Renn, 2019). Increasing polarization, for example between urgency and skepticism, expert knowledge and everyday experience, and individual claims to freedom and collective obligations, indicates that sustainable development cannot be achieved through scientific evidence or political programs alone. Rather, it is about the ways in which people describe themselves and the world (see, among others, Herdlitschka, 2025; Marg, 2023; Vogelmann, 2023). Taking this into account, the view of sustainable development shifts: normativity is deliberately set asidein favor of approaches that examine how societies generate meanings, how they interpret conflicts, and how they describe responsibility. But it also recedes in terms of how they relate the diversity of individual and collective lifestyles to one another. This is the central theme of our book, in which we pursue the goal of re-describing sustainable development in a neo-pragmatic way: as an open, fallibilistic process oriented toward human forms of life and experience, which is not stabilized by stabilizing truth claims, but rather by the quality of the procedures, vocabularies, and resonant relationships it produces.

O. Kühne et al., *Sustainable Development and Life Chances*, Neopragmatic Horizontal Geographies 3, https://doi.org/10.1007/978-3-032-20182-9_1

The approach developed in this book ties in with two theoretical strands, the combination of which opens up an independent perspective on sustainable development:

1. Neopragmatism, particularly as expressed by Richard Rorty (especially Rorty, 1979, 1989, 2023b). This makes it clear that social communication processes cannot be secured by objective criteria. Instead, they are secured by the way people describe each other and the world. These descriptions are contingent, historically situated, and always subject to revision. They are produced in open discussions. These discussions, in turn, depend on recognition, acceptability, and the renunciation of humiliation. In addition to extensive primary literature (among many: Brandom, 2001, 2008, 2011; Putnam, 1995a, 1995b; Rorty, 1979, 1997 [1989], 2023a; Shusterman, 2022), there is also extensive secondary literature (Calder, 2003; Horster, 1991; Müller, 2021, 2023; Reese-Schäfer, 2016) that is also available.[1] From a neo-pragmatic perspective, sustainable development appears to be a practice of redescription. In this practice, new vocabularies are created without violating the existing self-relationships of those involved.
2. The life chances perspective, which can be traced back in particular to Ralf Dahrendorf (especially Dahrendorf, 1979). This emphasizes how successful social orders and individual development are determined by the relationship between options and ligatures. Life chances arise on the one hand from the leeway that people need for their self-designs and on the other hand from voluntary bonds that enable meaning, orientation, and social belonging (further reading on Dahrendorf's approach: Kühne, 2017, 2025; Kühne & Leonardi, 2020; Peisert & Zapf, 1994). In this sense, we do not understand sustainable development as the implementation of a utopia or as a balance between diverging interests. Instead, we understand it as an enabling structure that combines individual self-creation and collective solidarity.

By combining these two strands, we develop an understanding of sustainable development that does not rely on stabilizing claims to truth, technocratic planning logic, or moral communication. Instead, we focus on the question of the conditions for successful transformation processes. The approach developed in this book understands sustainable development as a co-evolutionary relationship between individual and collective descriptions. An approach that is supported by compassion, recognition, and a culture of resonant self- and world relationships.

Our book thus ties in with Karl Popper's concept of the 'open society' (in which social development is shaped not by teleological designs but by fallibilistic learning processes). 'Openness' means a willingness to revise views or plans, to allow for alternatives, and to understand conflicts as productive moments of collective self-understanding. 'Closed societies', which are often characterized by utopian orientations, on the other hand, narrow the space for possible futures by reducing diversity

[1] In this respect, our book will limit itself to presenting key aspects of neopragmatism that are relevant to understanding the further explanations; however, it does not constitute a systematic introduction to neopragmatism.

and hindering or even preventing criticism (cf. Berr et al., 2025; Kühne et al., 2023). For sustainable development as an open, dialogical, and always contingent process, as we conceive it in this book, an open society is a necessary condition, because only in such a society can life chances grow, options be tried out, and bonds be reflected upon without hindering individual self-designs or understandings capable of accommodating plurality.

The book is divided into ten chapters. Chapter 1 introduces the initial problem and explains why a conceptual and normative redefinition of sustainable development is worthwhile. Chapter 2 provides the "analytical tools" that we use repeatedly throughout the book. Here we introduce the three-world theory and different concepts and traditions of criticism. In Chapter 3, we outline neopragmatism as the theoretical-conceptual framework for further discussion. Using vocabulary theory and an understanding of private self-creation and public solidarity, as well as the figure of the liberal ironist, we show how a non-essentialist perspective on sustainable development can be structured. The following Chap. 4 deepens this approach by examining the relationship between the individual and society through the concept of life chances. Here, it becomes clear how options and ligatures determine the structure of sustainable action. Chapter 5 first summarizes the theoretical background and derives criteria for further developing the vocabulary of sustainable development. Chapter 6 offers a historical and systematic review of the model of sustainable development and the Sustainable Development Goals. It becomes clear how various conceptual approaches have shaped the development of the understanding of sustainable development. In Chap. 7, we systematize central forms of criticism of the SDGs, as we have developed them in Kühne et al. (2024). This systematization ranges from internally oriented to ideological and metatheoretical to pragmatic perspectives. In addition, we show their significance for a conceptual reorientation. In the following Chap. 8, we link the life chances approach with sustainable development and present fundamental considerations and different positions on procedural understandings. These range from efficiency strategies to forms of weak sustainability. Chapter 9 forms the theoretical core of the book: here, a life chances-oriented understanding of sustainable development is re-described in the light of neopragmatism. In doing so, we link epistemological, ethical, and social-theoretical perspectives. These range from a departure from representationalism to questions of procedural and opportunity justice, which are related to compassion. Chapter 10 concludes the book with a summary. In addition to synthesizing the results, it outlines perspectives for open and resonant sustainable development.

References

Berr, K., Kühne, O., & Jenal, C. (2025). Landscape and Open Society. In O. Kühne, F. Weber, K. Berr, & C. Jenal (Eds.), *Landscape Handbook: German Language Research Perspectives* (pp. 77–91). Springer VS.

Brandom, R. B. (2001). *Making It Explicit: Reasoning, Representing, and Discursive Commitment* (4. print). Harvard University Press.
Brandom, R. B. (2008). Towards an Analytical Pragmatism. *Philosophical Topics, 36*(2), 1–27.
Brandom, R. B. (2011). Vocabularies of Pragmatism: Synthesizing Naturalism and Historicism. In R. B. Brandom (Ed.), *Perspectives on Pragmatism: Classical, Recent, and Contemporary* (pp. 116–157). Harvard University Press. https://doi.org/10.4159/9780674270572-006
Calder, G. (2003). *Rorty: And Redescription.* Weidenfeld & Nicolson.
Dahrendorf, R. (1979). *Lebenschancen: Anläufe zur sozialen und politischen Theorie. Suhrkamp-Taschenbuch: Vol. 559.* Suhrkamp.
Herdlitschka, T. (2025). Von fossilen zu erneuerbaren Energielandschaften: Machtverhältnisse und (Un-)Gerechtigkeiten in energiewendebezogenen Transformationsprozessen in der brandenburgischen Lausitz. In T. Mölders, J. Dannenberg, T. Herdlitschka, M. Hülz, & K. Kapitza (Eds.), *Kritische Nachhaltigkeits- und Transformationsforschung. Gender – Macht – Energiewende: Potenziale der Geschlechterforschung im Kontext raumbezogener Transformationen* (pp. 185–204). transcript.
Hoeft, C., Messinger-Zimmer, S., & Zilles, J. (2017). Bürgerproteste in Zeiten der Energiewende: Ein Fazit in neun Thesen. In C. Hoeft, S. Messinger-Zimmer, & J. Zilles (Eds.), *Bürgerproteste in Zeiten der Energiewende: Lokale Konflikte um Windkraft, Stromtrassen und Fracking* (pp. 235–254). transcript.
Horster, D. (1991). *Richard Rorty zur Einführung* (1. Aufl.). *Zur Einführung: Vol. 68.* Junius.
Kamlage, J.-H., Warode, J., Reinermann, J., de Vries, N., & Trost, E. (2020). Von Konflikt und Dialog: Manifestationen der Energiewende in den Transformationsfeldern Netzausbau, Biogas und Windkraft. In R. Duttmann, O. Kühne, & F. Weber (Eds.), *Landschaft als Prozess* (pp. 603–633). Springer VS.
Kühne, O. (2017). *Zur Aktualität von Ralf Dahrendorf: Einführung in sein Werk. Aktuelle und klassische Sozial- und Kulturwissenschaftler|innen.* Springer VS.
Kühne, O. (2025). Dahrendorf, Ralf (1929–2009). In M. Endreß & B. Rampp (Eds.), *Politische Soziologie* (pp. 315–322). Nomos.
Kühne, O., Berr, K., & Jenal, C. (2023). *The Closed Society and Its Ligatures—A Critique Using the Example of 'Landscape'.* Springer Fachmedien; Springer.
Kühne, O., & Leonardi, L. (2020). *Ralf Dahrendorf: Between Social Theory and Political Practice.* Palgrave Macmillan. https://doi.org/10.1007/978-3-030-44297-2
Kühne, O., Silina, I., & Berr, K. (2024). *Nachhaltige Entwicklungen und Lebenschancen.* Springer Fachmedien Wiesbaden. https://doi.org/10.1007/978-3-658-45998-7
Laborgne, P., & Radtke, J. (2023). Akzeptanz und Partizipation in der Energiewende. In M. Sonnberger, A. Bleicher, & M. Groß (Eds.), *Handbuch Umweltsoziologie* (pp. 1–15). Springer Fachmedien.
Lehmann, H. (2024). *Ideologiemaschinen: Wie Cancel Culture funktioniert.* Carl-Auer-Systeme Verlag.
Marg, S. (2023). Zwischen überhöhten Erwartungen und fehlendem Vertrauen: Wissenschaftsgläubigkeit und-skepsis im Kontext der Energiewende. *Demokratie Gegen Menschenfeindlichkeit, 8*(1), 30–45.
Mölders, T. (2025). Sozial-ökologische Perspektiven auf raumbezogene Transformationen der Energiewende: Entwicklung eines Forschungszugangs. In T. Mölders, J. Dannenberg, T. Herdlitschka, M. Hülz, & K. Kapitza (Eds.), *Kritische Nachhaltigkeits- und Transformationsforschung. Gender – Macht – Energiewende: Potenziale der Geschlechterforschung im Kontext raumbezogener Transformationen* (pp. 25–46). transcript.
Müller, M. (2021). *Rorty lesen.* Springer VS.
Müller, M. (2023). Einführung in das Handbuch Richard Rorty. In M. Müller (Ed.), *Handbuch Richard Rorty* (pp. 9–41). Springer Fachmedien.
Peisert, H., & Zapf, W. (Eds.). (1994). *Gesellschaft, Demokratie und Lebenschancen: Festschrift für Ralf Dahrendorf*. Deutsche Verlags-Anstalt DVA.

Popper, K. R. (1992a). *Die offene Gesellschaft und ihre Feinde: Der Zauber Platons* (7th ed., Vol. 1). J.C.B. Mohr. (Original work published 1945)

Popper, K. R. (1992b). *Die offene Gesellschaft und ihre Feinde: Falsche Propheten – Hegel Marx und die Folgen* (7th ed., Vol. 2). J.C.B. Mohr. (Original work published 1945)

Putnam, H. (1995a). *Pragmatism: An Open Question.* Blackwell.

Putnam, H. (1995b). Why There Isn't a Ready-made World. In P. K. Moser & J. D. Trout (Eds.), *Contemporary Materialism: A Reader* (pp. 233–253). Routledge.

Radtke, J. (2025). *How to Tackle Conflicts in Germany's Energiewende: On the Deeper Roots of Energy Justice Frictions and Disrupted Place Attachments.* https://papers.ssrn.com/sol3/papers.cfm?abstract_id=5262485

Radtke, J., & Renn, O. (2019). Partizipation und bürgerliche Engagement in der Energiewende. In J. Radtke & W. Canzler (Eds.), *Energiewende: Eine sozialwissenschaftliche Einführung* (pp. 283–316). Springer VS.

Reese-Schäfer, W. (2016). *Richard Rorty zur Einführung.* Junius.

Rorty, R. (1979). *Philosophy and the Mirror of Nature.* Princeton University Press.

Rorty, R. (1989). *Contingency, Irony, and Solidarity.* Cambridge University Press.

Rorty, R. (1997 [1989]). *Contingency, Irony, and Solidarity.* Cambridge University Press. https://doi.org/10.1017/CBO9780511804397

Rorty, R. (2023a). *Pragmatismus als Antiautoritarismus* (J. Schulte, Trans., & E. Mendieta, Ed.). Suhrkamp.

Rorty, R. (2023b). *Pragmatismus als Antiautoritarismus* (E. Mendieta, Ed.). Suhrkamp.

Shusterman, R. (2022). Pragmatist Philosophy for Our Times: Reviewing Rorty's Legacy. *Society, 59*(5), 583–590. https://doi.org/10.1007/s12115-022-00769-x

Stegemann, B. (2023). *Identitätspolitik.* Matthes & Seitz Berlin.

Vogelmann, F. (2023). *Umkämpfte Wissenschaften – zwischen Idealisierung und Verachtung.* Reclam Verlag.

Chapter 2
Analytical Tools: Three-World Theory and Understandings of Criticism

This chapter presents fundamental understandings that form an analytical basis for the further explanations. This initially concerns Karl Popper's Three Worlds Theory, which facilitates access to sustainable development by separating the levels of the physical, individual consciousness, and culture as a social construct. We then deal with the critical framework that we apply to current understandings of sustainability.

2.1 Karl Popper's Three Worlds Theory and Its Derivatives

Since criticism originally means differentiation, Karl Popper's three-world theory (cf. Kühne, 2020; Niemann, 2019; Popper, 1979, 2019 [1987]) offers a suitable model for analysis. It allows for a differentiated view of the relationship between the individual, the environment, and social constructs—also regarding the effects of the SDGs. As a quasi-ontological instrument, it makes it clear that human references to the world are not ontically separate but should be understood as different perspectives on a common reality (cf. Berr, 2020; Gethmann, 2023). *World 1* encompasses physical realities, *World 2* individual consciousness, and *World 3* cultural and scientific products (Popper, 2019 [1987], p. 17). A phenomenon such as a retention basin can thus be simultaneously physical (World 1), individually experienced (World 2), and culturally interpreted (World 3). This enables a flexible approach that does not ignore interactions but breaks them down analytically. In addition, Davidson (2004) distinguishes between three fundamental world references: *factual reference, social reference, and* self-reference—comparable in content to Habermas' objective, social and subjective world (Habermas, 1995). Davidson's concept of *intersubjectivity* shows that communication between worlds of consciousness (world 2) is achieved through language and shared reference objects (world 1). Through *triangulation* (Davidson, 2004), *radical interpretation*, and the *principle of charity* (Davidson, 2017), this understanding is possible despite individual differences. If this model is transferred

O. Kühne et al., *Sustainable Development and Life Chances*, Neopragmatic Horizontal Geographies 3, https://doi.org/10.1007/978-3-032-20182-9_2

to spatial and landscape contexts, space 1 is understood as physically given, space 2 as individually experienced, and space 3 as culturally constructed. The situation is similar with *landscapes 1–3* (cf. Kühne, 2024b). Socially shared interpretations (landscape 3) are transferred into individual consciousness (landscape 2) through socialization but can also flow back into landscape 3 from landscape 2 through innovations. These approaches are closely related to different *modes of perception* (Kühne, 2024a): *Mode a* is based on childhood images of home, *mode b* on everyday theoretical and media-mediated patterns and *mode c* on professional expertise. Depending on the perspective, not only the perception but also the normative evaluation of landscape varies—e.g., between nature conservation and energy production (Luhmann, 2017 [2002]).

2.2 Critical Framework

A well-founded critique of the Sustainable Development Goals (SDGs) requires systematic reflection on theoretical, methodological, and normative prerequisites. The aim of this discussion is not only to point out the shortcomings and challenges of the SDG framework but also to clarify the conditions and forms of legitimate scientific criticism (cf. Kühne et al., 2023).

2.2.1 Concept and Tradition of Criticism

The term 'criticism' (Greek kritiké techné) refers to the kind of judgment and discernment, as already established by Plato and Aristotle (Mittelstraß, 2004, pp. 498–499). Immanuel Kant transformed this understanding in the course of his epistemological shift toward self-reflection of reason. For Kant, criticism means not only the examination of empirical statements but also the limitation of the validity claims of human knowledge through an analysis of the conditions of its possibility (Kant, 1959 [1781], B 24). This approach implies a double movement: the "destruction of unfounded orientations and the construction of well-founded orientations" (Mittelstraß, 2004, p. 499).

In contrast to critical theory as represented by Horkheimer (Horkheimer, 1977, 1937]) or Foucault (Foucault, 2012 [1985]), the position adopted here pursues a separation of science and normative politics. In this context, criticism should remain primarily knowledge-oriented and reflexive. This follows the tradition of Popper and Dahrendorf, who emphasize an open, error-correcting scientific culture (cf. Dahrendorf, 1994; Kühne & Berr, 2021; Popper, 1973).

2.2.2 Typology of Scientific Criticism

In scientific discourse, six basic forms of criticism can be distinguished, which can be systematized along the poles of abstraction vs. concretion and internal vs. external scientific perspectives (cf. Kühne et al., 2023, pp. 39–41):

- *Internal criticism* (cf. e.g., Boltanski, 2010; Jaeggi, 2009; Stahl, 2013; Stederoth, 2011) focuses on the internal coherence of a research process. The central question is whether theoretical concepts, terms, methodological decisions, and methods have been appropriately chosen, clearly formulated, and applied in a context-appropriate manner. This form of criticism is usually expressed within the professional community but can also be taken up by external actors.
- *Criticism based on contextualization* analyzes scientific work in relation to comparable research projects. It examines whether the choice of theory, methods, and data is consistent with the existing field of research in terms of content and discourse (Kühne et al., 2023). The use of relevant literature is a central prerequisite for the connectivity of scientific knowledge.
- *Metatheoretical criticism* examines the theoretical foundations and epistemological assumptions of a research approach. It evaluates its potential to expand or limit existing theoretical frameworks. This particularly abstract, internally anchored criticism must be able to justify its standards theoretically (Kühne et al., 2023).
- *Ideological criticism* usually originates from non-scientific contexts and is based on normative assumptions—for example, of a religious, political, or moral nature (in the sense of Weber, 1976 [1922], 1988). Such criticism is often not very differentiated, as it does not explicitly state or systematically reflect on its normative foundations.
- *Criticism of the consequences and unintended side effects of research* (Berger, 2017 [1963]; Dahrendorf, 1968; Popper, 1963, 1945/1992) focuses on the unintended effects and ethical implications of research results. From a perspective of ethical responsibility (cf. Weber, 1976 [1922]), it asks to what extent scientific results are socially acceptable. This criticism can be formulated at various levels of abstraction, both internally and externally.
- *Pragmatic criticism* is strongly oriented toward concrete everyday problems and examines the practical relevance of research for the lifeworld. It asks whether scientific findings can contribute to solving practical challenges in life. This criticism is based on subjective experience (relevant: Gethmann, 1991; Gethmann et al., 2011; Husserl, 1954, 2008; Janich, 2011; Mittelstraß, 1991; Schütz & Luckmann, 2003, 1975; Waldenfels, 2005; Welter, 1986).

These categories are not mutually exclusive. Rather, they can overlap or be applied simultaneously to one and the same object of investigation. Their interaction opens up a differentiated understanding of criticism as a multidimensional scientific tool.

References

Berger, P. L. (2017 [1963]). *Einladung zur Soziologie: Eine humanistische Perspektive* (2., ergänzte Auflage). UVK; UVK/Lucius. (Original work published 1963)

Berr, K. (2020). Vom Wahren, Schönen und Guten: Philosophische Zugänge zu Landschaftsprozessen. In R. Duttmann, O. Kühne, & F. Weber (Eds.), *Landschaft als Prozess* (pp. 83–117). Springer VS.

Boltanski, L. (2010). *Soziologie und Sozialkritik*. Frankfurter Adorno-Vorlesungen 2008. Suhrkamp.

Dahrendorf, R. (1968). *Pfade aus Utopia: Arbeiten zur Theorie und Methode der Soziologie*. Piper.

Dahrendorf, R. (1994). *Der moderne soziale Konflikt: Essay zur Politik der Freiheit*. dtv.

Davidson, D. (2004). *Subjektiv, intersubjektiv, objektiv*. Suhrkamp.

Davidson, D. (2017). *Wahrheit und Interpretation* (6th ed.). Suhrkamp.

Foucault, M. (2012 [1985]). *Discipline and Punish: The Birth of the Prison*. Knopf Doubleday Publishing Group.

Gethmann, C. F. (Ed.). (1991). *Neuzeit und Gegenwart: Vol. 1. Lebenswelt und Wissenschaft: Studien zum Verhältnis von Phänomenologie und Wissenschaftstheorie*. Bouvier.

Gethmann, C. F. (2023). *Konstruktive Ethik: Einführung und Grundlegung*. Springer.

Gethmann, C. F., Bottek, J. C., & Hiekel, S. (Eds.). (2011). *Lebenswelt und Wissenschaft: XXI. Deutscher Kongreß für Philosophie 15. – 19. September 2008 an der Universität Duisburg – Essen*. Kolloquienbeiträge. Felix Meiner Verlag.

Habermas, J. (1995). *Theorie des kommunikativen Handelns: Band I. Handlungsrationalität und gesellschaftliche Rationalisierung*. Suhrkamp.

Horkheimer, M. (1977 [1937]). *Traditionelle und kritische Theorie: Fünf Aufsätze*. Fischer.

Husserl, E. (1954). *Die Krisis der europäischen Wissenschaften und die transzendentale Phänomenologie: Eine Einleitung in die phänomenologische Philosophie*. Herausgegeben von Walter Biemel. *Husserliana: Vol. 6*. Martinus Nijhoff.

Husserl, E. (2008). *Die Lebenswelt: Auslegungen der vorgegebenen Welt und ihrer Konstitution*. Texte aus dem Nachlass (1916–1937). Springer. https://doi.org/10.1007/978-1-4020-6477-7

Jaeggi, R. (2009). Was ist Ideologiekritik? In R. Jaeggi & T. Wesche (Eds.), *Was ist Kritik?* (pp. 266–298). Suhrkamp.

Janich, P. (2011). Handwerk und Mundwerk: Lebenswelt als Ursprung wissenschaftlicher Rationalität. In C. F. Gethmann, J. C. Bottek, & S. Hiekel (Eds.), *Lebenswelt und Wissenschaft: XXI. Deutscher Kongreß für Philosophie 15. – 19. September 2008 an der Universität Duisburg – Essen. Kolloquienbeiträge* (pp. 678–691). Felix Meiner Verlag.

Kant, I. (1959 [1781]). *Kritik der reinen Vernunft*. Felix Meiner Verlag.

Kühne, O. (2020). Landscape Conflicts: A Theoretical Approach Based on the Three Worlds Theory of Karl Popper and the Conflict Theory of Ralf Dahrendorf, Illustrated by the Example of the Energy System Transformation in Germany. *Sustainability: Science, Practice and Policy*, *12*(17), Article 6772, 1–20. https://doi.org/10.3390/su12176772

Kühne, O. (2024a). Sozialisation und Landschaft. In O. Kühne, F. Weber, K. Berr, & C. Jenal (Eds.), *Handbuch Landschaft* (2nd ed., 595–607). Springer VS.

Kühne, O. (2024b). Die Theorie der Drei Landschaften. In O. Kühne, F. Weber, K. Berr, & C. Jenal (Eds.), *Handbuch Landschaft* (2nd ed., pp. 251–260). Springer VS.

Kühne, O., & Berr, K. (2021). *Wissenschaft, Raum, Gesellschaft: Eine Einführung zur sozialen Erzeugung von Wissen*. Springer VS.

Kühne, O., Berr, K., & Lohmann, P. (2023). *Landschaft zwischen Philosophie und Sozialwissenschaften: Eine Kritik*. Springer Fachmedien and Springer VS.

Luhmann, N. (2017 [2002]). *Einführung in die Systemtheorie* (D. Baecker, Ed.). Carl-Auer Verlag.

Mittelstraß, J. (1991). Das lebensweltliche Apriori: Paul Lorenzen zum 70. Geburtstag. In C. F. Gethmann (Ed.), *Lebenswelt und Wissenschaft* (pp. 114–142). Königshausen & Neumann.

Mittelstraß, J. (Ed.). (2004). *Enzyklopädie Philosophie und Wissenschaftstheorie* (unveränderte Sonderausgabe, 2, H-O). J.B. Metzler.

Niemann, H.-J. (2019). Karl Poppers Spätwerk und seine ‚Welt 3'. In G. Franco (Ed.), *Handbuch Karl Popper* (Living Reference work, pp. 1–18). Springer Reference Geisteswissenschaften.

Popper, K. R. (1963). *Conjectures and Refutations: The Growth of Scientific Knowledge*. Routledge & Kegan.

Popper, K. R. (1973). *Objektive Erkenntnis: Ein evolutionärer Entwurf*. Hoffmann und Campe.

Popper, K. R. (1979). Three Worlds: Tanner Lecture, Michigan, April 7, 1978. *Michigan Quarterly Review, 18*(1), 141–167.

Popper, K. R. (1992). *Die offene Gesellschaft und ihre Feinde: Falsche Propheten – Hegel Marx und die Folgen* (7th ed., Vol. 2). J.C.B. Mohr. (Original work published 1945)

Popper, K. R. (2019 [1987]). *Auf der Suche nach einer besseren Welt: Vorträge und Aufsätze aus dreißig Jahren*. Piper.

Schütz, A., & Luckmann, T. (2003 [1975]). *Strukturen der Lebenswelt*. UTB.

Stahl, T. (2013). *Immanente Kritik: Elemente einer Theorie sozialer Praktiken*. Campus.

Stederoth, D. (2011). Kritik. In P. Kolmer & A. G. Wildfeuer (Eds.), *Neues Handbuch philosophischer Grundbegriffe* (Vol. 2, pp. 1346–1357). Alber.

Waldenfels, B. (2005). *In den Netzen der Lebenswelt* (3rd ed.). Suhrkamp.

Weber, M. (1976 [1922]). *Wirtschaft und Gesellschaft. Grundriß der verstehenden Soziologie*. Mohr Siebeck.

Weber, M. (1988). *Gesammelte Politische Schriften von Max Weber* (J. Winckelmann, Ed.) (5th ed.). Mohr Siebeck.

Welter, R. (1986). *Der Begriff der Lebenswelt: Theorien vortheoretischer Erfahrungswelt. Übergänge: Vol. 14*. Fink.

Chapter 3
Neopragmatism as a Theoretical-Conceptual Framework

Richard Rorty (2023) succinctly defines neopragmatism as a combination of American pragmatism and continental European philosophy of language. Pragmatism, which has its roots in the late nineteenth and early twentieth centuries, is associated in particular with Charles Sanders Peirce (1839–1914), William James (1842–1910), George Herbert Mead (1863–1931), and John Dewey (1859–1952). It was and is "not a unified school, but a pluralistic movement connected by a family feud. There is not one but many (neo-)pragmatisms" (Müller, 2023, p. 11). The term 'pragmatism' refers to 'action' in its etymological origin; Joas (2016 [1992], p. 28) summarizes it succinctly: "Pragmatism is a philosophy of action." Philosophical concepts are derived regarding practical consequences and must prove themselves by this standard: "For pragmatists, the meaning and relevance of philosophical concepts only becomes apparent in terms of practical consequences and their embedding in experiential, planning, and communicative processes of action" (Festl, 2018, p. 44). With the connection to the philosophy of language, the focus shifts from experience to conversation, the communicative negotiation of the world, even though the solution of concrete problems remains the goal of philosophical (or more generally, scientific) engagement with the world (among many: Kertscher, 2023; Müller, 2023).

The aim of this chapter is to present central aspects of neopragmatism (especially as defined by Richard Rorty) as a basis for our further considerations regarding sustainable development. We will first deal with the fundamentals of the neopragmatic philosophy of language, i.e., vocabulary theory. This, in turn, forms the basis for understanding neopragmatism as a metatheory. We will then look at the differentiation between the private and the public spheres and the question of public solidarity as a means of avoiding pain and an inclusive expansion of the 'we'. Finally, we will present Rorty's idea of the liberal ironist as a person who is aware of the contingency of self, language, and society.

O. Kühne et al., *Sustainable Development and Life Chances*, Neopragmatic Horizontal Geographies 3, https://doi.org/10.1007/978-3-032-20182-9_3

3.1 Vocabulary Theory

Conceptions of universal truth and irrefutable objectivity, against which hypotheses and theories can be clearly tested (Rorty, 1997 [1989]), cannot be upheld from a neopragmatic perspective. This also applies to the idea of language as a 'mirror of reality' or 'as the medium of knowledge and also the origin of a priori knowledge' (Müller, 2014, p. 63). The rejection of representationalist ideas does not mean the denial of a 'world out there': 'Truth cannot be out there—cannot exist independently of the human mind—because sentences cannot exist or be out there in that way. The world is out there, but descriptions of the world are not. Only descriptions of the world can be true or false. The world for itself—without the support of descriptive activity by humans—cannot exist" (Rorty, 1989, p. 24). We are connected to the 'world out there' "not in a representational way, but in a causal way" (Heindl, 2023, p. 10). Representationalist notions of 'truth' are replaced by the notion of 'more or less useful descriptions of the world and ourselves' (Rorty, 2018 [1994], p. 16). The idea of 'usefulness' can be summarized succinctly: "Useful for creating a better future" (Rorty, 2018 [1994], p. 16). These descriptions are expressed in vocabularies (Rorty, 1989), which can be understood as limited contexts of meaning; therefore, these vocabularies existed like separate chambers in an imaginary space (MacInyre, 2011). A vocabulary can be understood as "a bundle of metaphors that we have made our own" (MacInyre, 2011, p. 47). If one chooses a phenomenological reading, Rorty's vocabularies appear merely as relational descriptions of experiences within different contexts" (Weber, 2023, p. 7). The development of different vocabularies can be understood as a "phenomenon of linguistic division of labor" (Putnam, 1990, p. 43), a consequence of the complexity of the world. Vocabularies shape "mutually distrustful (and *not*: incomprehensible) communities of justification" (Rorty, 2023, p. 178). Vocabularies thus enable speakers to "draw on a network of beliefs that emerges in the communicative process of agreement and disagreement, justification and correction of beliefs, rather than through introspection that encounters an apparently pre-social foundation of knowledge in the subject" (Kronenberg, 2014, pp. 46–47). This creates a relational linguistic structure in which "words derive their meaning from their use in connection with other words" (Buschmeier, 2023, p. 3). This highlights the contingency of language, whereby Rorty's understanding of language is in turn linked to pragmatism: "For Rorty, language is a tool that we actively reshape as our needs, interests, and ideals evolve. Rorty refers to this process of adaptation as 'redescription'" (Huetter-Almerigi & Ramberg, 2023, p. 350).

Vocabularies are therefore always context-dependent. Contexts change, and vocabularies lose their claim to usefulness in describing the world or solving problems (Rorty, 1997 [1989]). Vocabularies thus become 'time-honored'. Through redescription, new vocabularies can emerge from these 'time-honored' vocabularies, which (following Dewey) means 'reconciling old innovations with newer ones' (Rorty, 2023, p. 252). Redescription is constitutively linked to the use of metaphors as "unfamiliar uses of old words" (Rorty, 1989, p. 80). Vocabularies can be further developed, whereby their use "is only possible against the background of other old words that continue to be used in familiar ways" (Rorty, 1989, p. 80). Redescriptions do not simply replace an old vocabulary with a new one. This would not be possible due to the constraints of language. New vocabularies can be understood as a 'fabric' in which 'time-honored' vocabularies are integrated into new ones (Topper, 1995). In this way, a new fabric recycles the remnants of old vocabularies (Rorty, 1997 [1989]; Topper, 1995). Redescriptions can also encompass several vocabularies, because "it is typical of revolutionary achievements in the arts, sciences, and moral and political thought that they can come about when someone recognizes interference between two or more of our vocabularies and then proceeds to invent a new vocabulary that can replace both" (Rorty, 2021 [1989], p. 35). Redescription "is a relentless attempt to express something new in the words of others while recognizing that what is said can only ever be parasitic" (Tautz, 2023, p. 10). Redescription, in turn, is often not a painless process, because "the most effective way to inflict lasting pain on people is to humiliate them by making everything that seemed particularly important to them appear futile, outdated, and powerless" (Rorty, 2021 [1989], p. 153).

This understanding of the vocabulary-based creation of the world can in turn be applied to the development of science.

3.2 (Meta-)theoretical-Epistemological Approach

If neopragmatism (here in the Rortyan sense) is made available for a scientific approach to the world, it can contribute, first, to ascertaining what is (scientifically) describable; second, to understanding science as an evolution of vocabularies; and third, to determining how the redescription of scientific vocabularies should develop—from a neopragmatic perspective, of course.

The rejection of a representationalist understanding and the search for the 'essence' of things are not part of a neo-pragmatic understanding of science: "Truth cannot be out there—it cannot exist independently of the human mind—because sentences cannot exist or be out there in that way. The world is out there, but descriptions of the world are not. Only descriptions of the world can be true or false. The world itself—without the support of human descriptive activity—cannot be" (Rorty, 1989, p. 24). An immediate understanding of the 'world' therefore remains closed to us. Consequently, the natural sciences, like other sciences, merely provide a contingent description of the world. The quality of descriptions of the world cannot be measured by a standard of 'consistency with reality', but rather by whether they prove useful in solving problems. The understanding of problems differs significantly from other philosophies of language (due to its pragmatic roots), as becomes clear when compared to poststructuralist deconstruction. Thus, it is clear "that Derrida likes to question things, whereas Dewey [following Rorty; authors' note] asks, 'What is the problem?' Our attitude [of Rorty and Dewey; authors' note] is: If something isn't broken, don't fix it. Keep using it until you find another tool that is more suitable. Derridarians tend to believe that the more questions, problematizations, and *mettant-en-abyme* they bring into everyday life, the better" (Rorty, 1999b, p. 104; emphasis in original). From a (neo-)pragmatist perspective, on the other hand, problems arise from the non-scientific world and not from the world of science itself. It is not old (scientific) questions that need to be asked anew, but questions that deal with the challenges of the present (Dewey, 1920). This in turn means that the use of neopragmatism as a (meta-)theory—like all theory—requires justification. After all, "the pressure to move to a higher level of abstraction should come from below" (Rorty, 1999a, p. 159), since (meta-)theoretical approaches are tools: "You only bring them out when you need to solve a specific problem" (Rorty, 1999a, p. 166). Thus, for Rorty, the ability to solve problems (Rorty, 1999a, p. 166) is the sole criterion for evaluating theories (of varying degrees of abstraction): "There is no criterion of inadequacy of a theory apart from the specification of such problems." In short, it can be formulated as follows: It is not the everyday world that needs justification, but abstractions.

As already explained, from a neo-pragmatic perspective, scientific theories are understood as tools. Tools that must always prove their usefulness in relation to solving problems. If they no longer meet this requirement, they are either sedimented or become the subject of a redescription. The need to use the tool of 'theory' arises for various reasons: firstly, because there is a wealth of empirical findings on a topic, space, phenomenon, etc. Secondly, because there are different (including theoretical) perspectives on the subject. Thirdly, because of the need to understand and relate descriptive, analytical, and normative statements about the world to one another. Particularly when there are numerous theories that illuminate a subject from different perspectives, meta-theoretical integration is necessary to be able to assess the relationships between the different theories (also understandable as vocabularies), to understand whether there are any incommensurabilities, or also to understand the complementarity between theories (Berr & Kühne, 2023). But also—and this is where Richard Rorty's concept of vocabularies and their redescription becomes relevant—to what extent they can be subjected to a common redescription. The extent to which neopragmatic redescriptions prove useful can be determined based on three criteria (Kühne, 2024; Kühne et al., 2024, 2025):

1. At the time of their creation (!), neopragmatic redescriptions must be more suitable for interpreting the 'world' and solving problems than 'traditional' vocabularies. This can also refer to objects that have newly emerged as a result of social or vocabulary development.
2. They must be designed in such a way that they can be integrated into the fabric of future redescriptions by being aware of their contextual dependence, temporally, socially, spatially, and theoretically. Accordingly, they do not formulate any exclusive 'claim to truth' (such as essentialist approaches to the world).
3. They enable a contingent interpretation of the world in two dimensions: on the one hand, in relation to dealing with the contingency of the world, and on the other hand, in relation to the production of contingent interpretations of the world.

A central aspect of the idea of redescribing theoretical vocabularies is that, with a focus on utility, questions lose their significance, and redescriptions are not aimed at eliminating vocabularies, but rather at constructive reuse and complementarity: "I will not put forward arguments against the vocabularies I wish to replace. Instead, I will try to make my preferred vocabularies attractive by showing how they can re-describe various topics" (Rorty, 1997 [1989], p. 9). In this way, results can be integrated, even theoretical aspects that run counter to the post-representationalist orientation of neopragmatism, such as positivism. Not because it has succeeded in truthfully depicting the world, but because it has succeeded in generating knowledge about the world that has proven useful, for example, in relation to a cartographic recording of the world, even if this is a social construction based on a conventionalization that is not power-neutral (see Edler & Kühne, 2022; Edler et al., 2023; Kühne & Edler, 2025).

Redescriptions of scientific vocabularies, in turn, pose challenges: The first has already become clear: even if a redescription causes less pain than attempting to replace one vocabulary with another, supporters of a redescribed vocabulary will not necessarily be willing to acknowledge the greater usefulness of the redescribed vocabulary without objection. A redescription, in turn, is based on an extensive examination of different scientific vocabularies: "A successful 'redescription' requires considerable knowledge, insight, and skill" (Malachowski, 2002, p. 183). Furthermore, when redescribing, the focus should remain on problem solving, in short, on designing triangulations that are appropriate to the subject and do not lead to over-complexification (keyword: Ockham's razor; Rodriguez-Fernández, 1999). The need for justification, which goes hand in hand with a neopragmatic approach, can be understood here as a tool to counteract over-complexification.

3.3 Private and Public Space

Anticipating the dual (*not* dualistic) distinction between individual pursuit of happiness and action on the one hand and societal normative expectations and bonds (ligatures) on the other (Sect. 4.1), the separation between private and public space can also be defined as a dual distinction. This separation is particularly linked to the demand "that the state should be kept within its limits and society protected from political encroachment. The boundaries of the private sphere are the boundaries of the political sphere. The defense of privacy is the individual's most effective objection to the fatal universalism of power" (Sofsky, 2007b, p. 23). In this sense, privacy represents the most important protection "against the constraints of roles, collectives, and institutions" (Sofsky, 2007a, p. 51) or social ties. The protection of privacy, and in particular of private property, begins with fencing off one's possessions and protecting them with walls: "The wall is one of humanity's most important inventions, on a par with the wheel, the plow, and writing. It creates distance and protects against encroachment. Behind the wall, individuals can lay down the weapons they use to arm themselves against the impositions of the public sphere. The wall secures personal freedom" (Sofsky, 2007b, p. 30). Attempts to abolish private property, on the other hand, have disastrous consequences for social and spatial development: "The abolition of property degrades the individual to a public figure.

In property, the will of the individual gains tangible reality" (Sofsky, 2007b, p. 99). Private property—especially ownership of houses and land—enables individuals to reduce social and communal complexity. It is based on respect for privacy. Citizens respect that "property remains hidden from the ears and eyes of the public" (Sofsky, 2007b, p. 100). Accordingly, freedom and private property are subject to a reciprocal relationship: "There is no personal freedom without political guarantees of private property, and there is no independence of lifestyles, opinions, and actions without a minimum of economic independence" (Sofsky, 2007b, p. 100). Private space 1 is thus a refuge from the demands (not accepted by space 2) of space 3. It is removed from the ongoing conflicts of public space, which are currently becoming virulent in the context of sustainability and transformation demands. The attempt to resolve associated conflicts in a collectivist abolition of the private sphere in a public space stifles the potential productivity of such conflicts and ultimately leads to agony for those involved.

Richard Rorty also uses a dual distinction in his neo-pragmatic metatheory that corresponds to the distinctions already mentioned: namely, that of 'private self-creation' and 'public solidarity' (2021 [1989]). The 'private space' refers to the realm of the private and self-creation; in this sense, Rorty speaks of a vocabulary of self-creation that is necessarily private and not public. The 'public space' refers to the realm of the public and solidarity. In this sense, Rorty speaks of a vocabulary of justice that is necessarily public and not private. This distinction is therefore embedded in Rorty's vocabulary theory, within the framework of which he concludes, "There is no way to bring self-creation and justice together on a theoretical level"; one can only "take practical measures to achieve this practical goal" (Rorty, 1989, p. 13). This is because the "vocabulary of self-creation is necessarily private, shared by no one, and unsuitable for argumentation. The vocabulary of justice is necessarily public, shared by many, a medium for the exchange of arguments" (Rorty, 1989, p. 13).

3.4 The Expansion of 'We' According to Richard Rorty

Vocabularies enable speakers to "draw on a network of beliefs that is formed in the communicative process of agreement and disagreement, justification and correction of beliefs, rather than through introspection that encounters an apparently pre-social foundation of knowledge in the subject" (Kronenberg, 2014, pp. 46–47). The emergence and use of vocabularies have a specific consequence: in and with vocabularies, "mutually distrustful (and *not*: incomprehensible) communities of justification" (Rorty, 2023, p. 178; emphasis in original) emerge. This formation of mutually distrustful communities is associated with exclusivist tendencies: "The other 'does not belong to us,' he does not belong to the *real* people, the *paradigmatic* people, that is, to those who are to be treated with respect in terms of their person and their opinions" (Rorty, 2023, pp. 178–179; emphasis in the original). This means that the formation of vocabularies always carries the risk of causing pain to other people, most effectively by humiliating them (Rorty, 1989). Fundamentally opposed to this is a basic moral attitude of compassion (Rorty, 1997 [1989], 1999c), on which a humanized culture is based. In this sense, it should be understood as "our moral task in the sense of expanding our moral community and ensuring, through inclusion, that more and more diverse people fall under our use of the word 'we'" (Rorty, 2023, p. 55).

3.5 Orientation Toward the Liberal Ironist: Individualization of Sustainable Development and Public Solidarity

Based on his insight into the triple contingency of language, self (which arises from self-creation in confrontation with vocabularies), and society (which also follows the evolutionary process of redefinition and has neither essence nor goal), Rorty advocates a fundamentally ironic attitude (1997 [1989]). This attitude is an expression of the insight that one's own vocabulary is also contingent. For Rorty, 'contingency' is "at one end of a set of oppositions whose other term is *variously necessary, essential, intrinsic*, or unconditional" (Topper, 1995, p. 958). This attitude and insight into contingency give rise to the figure of the 'liberal ironist' as a person "who faces the fact that her central beliefs and needs are contingent" (Rorty, 1989, p. 14): "Liberal ironists are people who count among these needs, which cannot be traced back to the deepest reasons, their own hopes, the hopes that suffering will diminish, that the humiliation of people by people will perhaps cease" (Rorty, 1989, p. 14).

Private self-creation does not take place in a space beyond other people, but on the contrary, precisely in the context of other people: "The place of freedom is not thinking or spiritual vision, but communicative life practice." (Müller, 2014, p. 411). This has an effect on the cohesive forces of a society. "The social glue that holds the […] ideal liberal society together consists only in the consensus that a social organization is about giving everyone a chance for self-creation—according to their possibilities—and that this goal requires, in addition to peace and prosperity, the usual 'civil' liberties" (Rorty, 2021 [1989], p. 145). In this sense, it can be argued that an 'ideal liberal society' is determined by specific consensually validated procedural rules within which people have the opportunity for self-creation, which they can use—or not (see Kühne, 2024; Kühne et al., 2021).

Applying these considerations on the relationship between private self-creation and public solidarity to the question of implementing sustainability goals yields the following guidelines: Since self-creation is always dependent on social conditions and is always located in a social context, social ligatures must not be imposed on individuals in an unnecessary, counterintuitive, self-serving, and undemocratic manner. However, in the case of democratically negotiated and widely accepted social goals, these goals and the means of achieving them can certainly be imposed on the individual. Conversely, private self-creation options must not destroy the 'social glue' and undermine the procedural rules that enable and guarantee the opportunities for self-creation. It should therefore be possible to individualize sustainable development in the course of private self-creation. It should also be possible to achieve sustainable development based on public solidarity that respects private self-creation but also sets a framework of strict rules of acceptibility for this individualization of sustainable development.

3.6 Summary

Neopragmatism—especially as defined by Richard Rorty—can be understood, as explained in this chapter, as a theoretical stance that opposes the claim of representationalist truth. Instead, it aims to expand contingency, particularly in problem solving, in and through vocabularies. Science is not understood here as the key to discovering eternal truths but as a bundle of cultural practices of description, redescription, and understanding. Like other vocabularies, scientific vocabularies must repeatedly prove their usefulness in solving problems. From this perspective, theory is understood as a tool. The questions asked here are not "What is?" but "What for?" or "What helps?" Vocabularies that have to constantly prove themselves change. One trigger for change is new metaphors to which a vocabulary must respond. A fundamental change in vocabularies, on the other hand, takes place in redescriptions, understood as the creative transformation of familiar descriptions. 'Time-honored' vocabularies are reworked here, either in whole or in part, into new vocabularies—without claiming to be the ultimate justification, but with the aim of achieving newly gained practical connectivity. However, redescription is also a painful process—after all, it often devalues familiar interpretations of the world or approaches to problem solving that many have not yet abandoned. In this respect, redescriptions also need to be justified—weighing up the pain they cause against the expectations regarding their problem-solving capabilities.

The distinction between private self-creation and public solidarity makes it possible to outline an ethic based on context-sensitive negotiation—not on concrete (ontological) universal utopias. The figure of the liberal ironist stands here for an attitude that recognizes the contingency of one's own convictions without falling into cynicism. Aware of the contingency of self, language, and society, she feels committed to the inclusive expansion of the 'we'. This expansion of the 'we', in turn, means a 'meta-utopia' of a humanized society based on a solidarity-based attitude of pain avoidance.

Against this background, neopragmatism can be understood as a metatheoretical framework in which sustainability is not conceived as a teleological goal but rather as an ongoing process of understanding and transformation, of constant reflection on the appropriateness of decisions made and one's own vocabulary constraints. This implies a process that integrates both individual life chances and collective responsibilities, as the two are inextricably linked. We will address these issues in more detail in the next chapter.

References

Berr, K., & Kühne, O. (2023). Aesthetic Constructions of Landscape Between Society, Individual and Objects: A Neopragmatic Approach. *Studi di Estetica, LI, IV Serie, 26*, 1–29. https://doi.org/10.7413/1825864638

Buschmeier, M. (2023). Sprachphilosophie: Sprachspielpragmatismus und -historismus. In M. Müller (Ed.), *Handbuch Richard Rorty* (pp. 597–614). Springer Fachmedien.

Dewey, J. (1920). *Reconstruction in Philosophy*. Henry Holt and Co. https://permalink.obvsg.at/

Edler, D., & Kühne, O. (2022). Aesthetics and Cartography: Post-Critical Reflections on Deviance in and of Representations. *ISPRS—International Journal of Geo-Information, 11*(10), Article 526. https://doi.org/10.3390/ijgi11100526

Edler, D., Kühne, O., Keil, J., Korte, A., Weißmann, M., Galla, C., Zilt, L., & Dickmann, F. (2023). Geodatenkultur der Diversität: eine neopragmatische Skizzierung. *KN—Journal of Cartography and Geographic Information, 73*, 317–333. https://doi.org/10.1007/s42489-023-00146-6

Festl, M. G. (Ed.). (2018). *Handbuch Pragmatismus*. J.B. Metzler. https://doi.org/10.1007/978-3-476-04557-7

Heindl, A. (2023). Richard Rortys Kritik der Erkenntnistheorie: Demokratischer Anti-Repräsentationalismus. In M. Müller (Ed.), *Handbuch Richard Rorty* (pp. 581–596). Springer Fachmedien.

Huetter-Almerigi, Y., & Ramberg, B. T. (2023). Analytic Philosophy of Language (Wittgenstein, Sellars, Quine, Davidson, Kuhn). In M. Müller (Ed.), *Handbuch Richard Rorty* (pp. 347–362). Springer Fachmedien.

Joas, H. (2016 [1992]). *Pragmatismus und Gesellschaftstheorie* (3. Aufl.). *Suhrkamp-Taschenbuch Wissenschaft: Vol. 1018*. Suhrkamp.

Kertscher, J. (2023). Rorty und der klassische Pragmatismus (Peirce, James, Dewey, Schiller). In M. Müller (Ed.), *Handbuch Richard Rorty* (pp. 333–346). Springer Fachmedien.

Kronenberg, B. (2014). *Die Zerbrechlichkeit des Wahren: Richard Rortys Neopragmatismus und Adornos Negative Dialektik* (1st ed.). *Edition Moderne Postmoderne*. transcript. https://doi.org/10.5555/9783839414101

Kühne, O. (2024). *Redescribing Horizontal Geographies: A Neopragmatist Approach to Spatial Contingency, Complexity, and Relationships*. Springer International.

Kühne, O., Berr, K., Edler, D., Lohmann, P., & Schuster, K. (2025). *Neopragmatismus – inverse Landschaft – (karto)graphische Darstellung: Von der Kritik zur Weiterentwicklung*. Springer VS.

Kühne, O., Berr, K., Schuster, K., & Jenal, C. (2021). *Freiheit und Landschaft: Auf der Suche nach Lebenschancen mit Ralf Dahrendorf*. Springer.

Kühne, O., & Edler, D. (2025). Reconstructing the Map: A Neopragmatist Perspective on Cartography in the Context of Artificial Intelligence (AI). *KN—Journal of Cartography and Geographic Information, 75*, 79–92. https://doi.org/10.1007/s42489-024-00184-8

Kühne, O., Koegst, L., & Berr, K. (2024). *Oilscapes of Louisiana—Neopragmatic Reflections on the Ambivalent Aesthetics of Landscape Constructions*. Springer Fachmedien.

MacInyre, A. (2011). Richard Rortys Vermächtnis. In A. Gröschner, M. Sandbothe, & R. Rorty (Eds.), *Suhrkamp Taschenbuch Wissenschaft: Vol. 1981. Pragmatismus als Kulturpolitik: Beiträge zum Werk Richard Rortys* (1. Aufl., Orig.-Ausg, pp. 39–53). Suhrkamp.

Malachowski, A. (2002). *Richard Rorty*. Acumen Publishing.

Müller, M. (2014). *Private Romantik, öffentlicher Pragmatismus? Richard Rortys Transformative Neubeschreibung des Liberalismus* (1st ed.). *Edition Moderne Postmoderne*. transcript. https://doi.org/10.5555/9783839420416

Müller, M. (2023). Einführung in das Handbuch Richard Rorty. In M. Müller (Ed.), *Handbuch Richard Rorty* (pp. 9–41). Springer Fachmedien.

Putnam, H. (1990). *Die Bedeutung von „Bedeutung“* (2., durchgesehene Auflage). Klostermann.

Rodriguez-Fernández, J. (1999). Ockham's Razor. *Endeavour, 23*(3), 121–125. https://doi.org/10.1016/S0160-9327(99)01199-0

Rorty, R. (1989). *Kontingenz, Ironie und Solidarität*. Suhrkamp.
Rorty, R. (1997 [1989]). *Contingency, Irony, and Solidarity*. Cambridge University Press. https://doi.org/10.1017/CBO9780511804397
Rorty, R. (1999a). Antwort auf Ernesto Laclau. In C. Mouffe (Ed.), *Dekonstruktion und Pragmatismus: Demokratie, Wahrheit und Vernunft* (pp. 155–170). Passagen-Verlag.
Rorty, R. (1999b). Antwort auf Simon Critchley. In C. Mouffe (Ed.), *Dekonstruktion und Pragmatismus: Demokratie, Wahrheit und Vernunft* (pp. 97–110). Passagen-Verlag.
Rorty, R. (1999c). *Philosophy and Social Hope*. Penguin Books.
Rorty, R. (2018 [1994]). *Hoffnung statt Erkenntnis: Eine Einführung in die pragmatische Philosophie* (3rd ed.). Passagen-Verlag.
Rorty, R. (2021 [1989]). *Kontingenz, Ironie und Solidarität* (13th ed.). Suhrkamp.
Rorty, R. (2023). *Pragmatismus als Antiautoritarismus* (J. Schulte, Trans.) (E. Mendieta, Ed.). Suhrkamp.
Sofsky, W. (2007a). Das Prinzip Freiheit. In U. Ackermann (Ed.), *Welche Freiheit: Plädoyers für eine offene Gesellschaft* (pp. 40–61). Matthes & Seitz.
Sofsky, W. (2007b). *Verteidigung des Privaten: Eine Streitschrift*. Beck.
Tautz, B. (2023). Klassiker der Ironie als Lebensform (Sokrates, Kierkegaard). In M. Müller (Ed.), *Handbuch Richard Rorty* (pp. 443–458). Springer Fachmedien.
Topper, K. (1995). Richard Rorty, Liberalism and the Politics of Redescription. *American Political Science Review, 89*(4), 954–965. https://doi.org/10.2307/2082520
Weber, B. (2023). Die liberale Ironikerin: Von der politischen Intellektuellen zur realistischen Träumerin. In M. Müller (Ed.), *Handbuch Richard Rorty* (pp. 883–899). Springer Fachmedien.

Chapter 4
Balancing the Individual and Society

The relationship between the individual and society is a topic of discussion in traditional philosophy, particularly social philosophy, as well as in sociology and political science. Put simply, there are two distinct approaches: those that start methodologically from the individual and ask how communities or societies can arise from the aggregation of individuals, and those that start methodologically from social structures and ask how the individual can be integrated into these structures. The conflictual nature of this relationship is undisputed, as is evident, for example, in the political science distinction between individualistic and collectivist approaches: Fundamentally, collectivist thinking attributes greater importance to communities or social groups than to individuals; the value of a person is defined by their fit with and integration into the community. Individualistic thinking reverses these weightings. In each case, this relationship is associated with a hierarchy and thus potential for conflict, or, in Kant's words, an "antagonism" (Kant, 1983 [1784], p. 37).

This 'antagonism' is nothing other than the "*unsociable sociability* of human beings," which is "apparently in human nature". On the one hand, human beings have a tendency to "socialize" because they feel the development of their abilities in a state of socialization. On the other hand, however, they also have "a great tendency to isolate *themselves*, because they simultaneously encounter within themselves the unsociable trait of wanting to judge everything solely according to their own sense, and therefore expect resistance everywhere, just as they know from themselves that they are inclined to resist others" (Kant, 1983 [1784], p. 38).

Kant thus addresses the antagonism between socialization ('sociability') and individualization ('unsociability'), as Ralf Dahrendorf, for example, has also taken up theoretically. Dahrendorf explicitly refers to the irrefutable and 'annoying fact' of 'society' (Dahrendorf, 1972), which can be traced back to the thesis already formulated by Aristotle (2009, 1253a) that man is a '*zôon* politikón', that is, a 'social being' (Dahrendorf, 1972, p. 282). Thus, humans can "only realize themselves through society. The social existence of humans is a condition for the possibility of their freedom. At the same time, however, their social existence is a condition for the

O. Kühne et al., *Sustainable Development and Life Chances*, Neopragmatic Horizontal Geographies 3, https://doi.org/10.1007/978-3-032-20182-9_4

possibility of their lack of freedom, for society always implies coercion and restriction. The act of socialization [...] is necessarily an act of submission to rules, norms, and controls" (Dahrendorf, 1972, p. 282; see also Dahrendorf 1957).

4.1 The Dual Structure of Individual Aspirations and Options and Social Expectations and Ties

The question is whether and how this antagonism between the individual and society, or between individualization ('unsociability') and socialization ('sociability'), can be balanced. To answer this question, it makes sense to refer to a distinction between two fundamental ethical models that is well known in philosophical ethics: that between 'ethics of striving' and 'ethics of ought' (Krämer, 1992) or 'eudaemonistic ethics' and 'normative ethics' (Kamlah, 1973). This reference can show that an antagonistic opposition between World 2 and World 3, i.e., between individual consciousness and individual interests, preferences, intentions, and options on the one hand, and social conventions, cultural symbols and interpretation systems, and ligatures on the other, makes little sense.

The 'I' that Dahrendorf focuses on is favored in what is known as 'ethics of striving', which since ancient times and the Middle Ages has explored the question of how, in the 'use' of freedom, an individual's relationship with themselves can enable them to lead and shape their own life in a self-determined way (Hadot, 2005; cf. Horn, 2014). Wilhelm Kamlah describes this tradition of a philosophically determined 'art of living' as 'eudaimonistic ethics,' which must pursue the question of 'how we can live' (Kamlah, 1973, pp. 145–182; cf. also Krämer, 1992). Hans Krämer speaks of 'ethics of striving', which consistently begins and ends its reflection on morality with the 'I want' of the individual and is therefore also referred to as 'first-person ethics' (Krämer, 1992, p. 84). Kamlah contrasts 'eudaimonistic ethics' with 'normative ethics', which is also identified with 'moral philosophy' and the question of 'how we should live' (Kamlah, 1973, pp. 93–144). Krämer contrasts 'aspirational ethics' with 'deontological ethics', which emphasizes the primacy of society's demands on the individual, which, from the perspective of the 'second and third person' (cf. Krämer, 1998, p. 97), leads reflection on the perspective of the individual to result in an 'I ought' instead of an 'I want'. Kamlah speaks of one type of ethics being the 'necessary complement' (Kamlah, 1973, p. 144) to the other, while Krämer emphasizes that it is necessary to "insist on the irreducibility of the dual structure of ethics," which "is therefore necessarily bicentric" (Krämer, 1992, p. 122) and clarifies Kamlah's basic idea of the 'necessary complementarity' to the statement that the two ethical perspectives are "complementary to each other and therefore cannot mutually cancel each other out or replace each other" (Krämer, 1992, p. 84).

Attempts at reduction are doomed to failure because the "gap between the common good and self-interest cannot be bridged in principle, and if it can, then only in an infinite approximation" (Krämer, 1992, p. 40). Kamlah therefore reconstructs a "new fundamental question" for both types of ethics (Kamlah, 1973, p. 144). Eudaimonistic ethics poses "the question of being able to live" as "the question of a 'successful life'" (Kamlah, 1973, p. 147); and this question gives rise to the "basic experience" and "basic insight" that "true ability to live [...] is first and foremost the ability to let go" and that every human being is "on a hopeless path in their self-will and self-consciousness" (Kamlah, 1973, p. 158)—from which the complementary consideration of social aspects arises. Normative ethics thus additionally asks "how we *should* live" (Kamlah, 1973, p. 94), from which the "practical basic norm" results, namely that "everyone is required at all times to recognize that their fellow human beings are as needy as they are, and to act accordingly" (Kamlah, 1973, p. 96)—which results in the complementary consideration of individual aspects. In Dahrendorf's vocabulary, this means that options require ligatures, just as ligatures require options.

4.2 The Concept of Life Chances

Max Weber introduces the social science term 'life chances' and understands it to mean the opportunities that a person can reasonably expect in their life (M. Weber, 1976 [1922]; see also Berger, 2017 [1963]). He outlines life chances on the one hand as "structurally based [...] probability of behavior" and, on the other hand, "as something that the individual can have, something as an opportunity to satisfy interests" (Dahrendorf, 1979, p. 98). Ralf Dahrendorf further differentiates Weber's concept. He also adds a normative component to it (Mackert, 2010). For Dahrendorf (1979, p. 24), life chances are—figuratively speaking—"the molds of human life in society; they determine the extent to which people can develop." He understands "life chances" to mean "first and foremost, choices and options. They require two things: rights to participation and a range of activities and goods to choose from" (Dahrendorf, 2007, p. 44). Opportunities for choice must be linked to meaning. Life chances arise from a productive relationship between the individual and society because they are "opportunities for individual growth, the realization of abilities, desires, and hopes, and these opportunities are provided by social conditions" (Dahrendorf, 1979, p. 50). However, society does not only enable life chances; it also restricts them. This becomes particularly clear in the case of class differences, when "the opportunities of some (the 'dominated', the 'dependent') are decisively pre-structured by the decisions of others ('the rulers')" (Niedenzu, 2001, p. 178). For Dahrendorf, the social context remains ambivalent (Dahrendorf,), because on the one hand it enables the pursuit and realization of life chances (or is absolutely necessary), while on the other hand it hinders the pursuit and realization of life chances.

Life chances arise from a relationship between options and ligatures (Dahrendorf, 1979, p. 50). Options are "choices given in social structures, alternatives for action" (Dahrendorf, 1979, p. 108). Ligatures are understood as "structurally predetermined fields of human action. Individuals are placed in bonds or ligatures by virtue of their social positions and roles". This highlights the ambivalent relationship between the individual and society: ligatures bind the individual to social contexts, while options are a prerequisite for self-empowerment and emancipation in and from these contexts. Ligatures form "in a sense the inside of the norms that guarantee social structures" (Ackermann, 2020, p. 141). They are emotionally and morally charged, producing obligation on the one hand and belonging on the other. Ligatures thus function as "foundations of action" (Dahrendorf, 1979, p. 51), such as "ancestors, homeland, community, church" (Dahrendorf, 1979, p. 51). Options and ligatures are characterized by a reciprocal relationship: "Ligatures without options mean oppression, while options without ties are meaningless" (Dahrendorf, 1979, pp. 51–52). Ligatures transform mere opportunities into "opportunities with meaning and significance, i.e., opportunities in life" (Dahrendorf, 1979, p. 51). They also have "an absolute quality: they are reluctant to allow shades of gray. People either belong or they don't, and if they don't, they have no claim to rights" (Dahrendorf, 1979, p. 55). Ligatures thus have a highly ambivalent effect on the development of life chances, while options always have a positive influence (Kühne & Leonardi, 2020; Kühne et al., 2021; Mackert, 2010). An increase in life chances is therefore justified on the one hand by a gain in options, which on the other hand is achieved by "breaking up ligatures" (Dahrendorf, 1979, p. 52). A social development geared towards life chances requires a "policy of options, i.e., the struggle to expand human life chances by increasing options" (Dahrendorf, 1983, pp. 123–124).

However, the maximization of life chances faces various obstacles (Bovone, 1982; Dahrendorf, 1968, 1979, 1994, 2006; Kocka, 2004; Kühne et al., 2021, 2022a):

1. A lack of individual ability to deal critically with ligatures, especially due to low levels of education, reduces the chance of emancipating oneself from outdated ligatures that are accepted unquestioningly and not chosen voluntarily.
2. The renunciation of developing options through 'external guidance' (Riesman, 1950) and the unquestioning adherence to social conventions (especially ligatures) lead to a reduction in life chances, even if the ability to criticize through education is present in principle.
3. Individual life chances can be restricted by bureaucratic regulation. This occurs at different levels: at the concrete level, by restricting options, and also at the level of time opportunity costs, since time that must be spent dealing with administrative ligatures can no longer be used to develop options. However, this also affects the choice of an exit option, i.e., no longer devoting oneself to the development of options, as dealing with formalities is considered too costly in relation to the expected gain (Kühne, 2025).
4. The systematic social inequality of life chances means reduced opportunities for the development of life chances for parts of society. This can occur in particular through the systematic discrimination of parts of the population in terms of access to education, but also through the conversion of economic power into political power.
5. The latter point in particular goes hand in hand with the conversion of options into ligatures for the purpose of securing privileges. In short, increasing options for the few goes hand in hand with the increased use of ligatures for the many.

6. Refusing to engage individually with the complexity of the relationship between options and ligatures by escaping into utopias leads to a reduction in life chances. Utopias, in turn, are often morally secured by ligatures.

The last point made it clear that ligatures can be presented in a very differentiated way. In different contexts, we have worked on differentiating the concept of ligatures recently to make the ambivalence of the concept more tangible on the one hand and to make the diversity of ligatures more comprehensible on the other (for details, see Kühne, 2024b; Kühne & Koegst, 2023; Kühne et al., 2021, 2022a, 2023). This differentiation could be made on four levels:

1. Moral and ethical ligatures. Moral ligatures form unquestioned norms of behavior (such as religious norms) that individuals are expected to follow. Ethical ligatures are tools for reflecting on moral ligatures, but they also offer options. While moral ligatures limit options, ethical ligatures can give meaning to options.
2. Internal and external ligatures. Following David Riesman's considerations (Riesman, 1950) on internal and external character, ligatures can be understood as internal, focused on one's own actions. External ligatures can be understood as those that focus on the actions of others. Internally directed ligatures thus set standards for one's own actions, while externally directed ligatures (attempt to) set standards for the actions of others.
3. Explicit and implicit ligatures. Explicit ligatures are clearly formulated, while implicit ligatures tacitly assume that they must be known to everyone. Explicit ligatures can be subjected to simple critical reflection, whereas implicit ligatures can only be subjected to such reflection if they are made explicit. Implicit ligatures also have great potential for distinction.
4. Physical and physically mediated social ligatures. Physical ligatures develop from the experience of human physicality. Physical ligatures arise from an awareness of the vulnerability of the human body, its limited physical capabilities, including its impact on the (rest of the) material world. Physically mediated social ligatures make human physicality the subject of social practices. These physically mediated social ligatures are diverse and range from the organization of maintaining the body's vital functions (the organization of food, warmth, protection from injury, etc.) to regulations governing coexistence (e.g., to avoid injury, for example, in road traffic) to the punishment of deviant behavior by restricting physically bound options (e.g., through imprisonment).

This differentiation of ligatures represents an analytical scheme to illustrate how ligatures work. In the social world, they commonly occur in various mutual interpenetrations: ethical ligatures are often explicit and thus form a basis for productive intersubjective interaction. Here it becomes clear how the (common) rules of reflection on ligatures and options are structured. Outward-looking morals and implicit ligatures often occur. Outward-looking moral ligatures compete for interpretation. Ultimately, the differentiation of society is also linked to a differentiation of moral ligatures. Moral communication also tends to devalue others while simultaneously elevating oneself (Luhmann, 1993, 2016). This gives rise to conflicts that are difficult or impossible to resolve (Berr et al., 2022; Kühne, 2019; Kühne et al., 2022b), as the position of the other party to the conflict is ultimately not recognized as equal (Dahrendorf, 1972, 1992). The option-limiting function of outward-oriented moral ligatures becomes even more pronounced when the ligatures remain implicit: instead of explaining moral ligatures, it is assumed that others share one's own moral norms—because these are taken for granted (for more concerning energy transition: Lohmann et al., 2024; Thiel & Kühne, 2024).

The concept of options can also be further differentiated. However, this differentiation relates less to the content (as in the differentiation of the ligature concept) and more to the structural characteristics (for more details, see Kühne, 2024a; Kühne et al., 2024, 2025; Thiel & Kühne, 2024):

1. Latent options. This describes the total number of options available in a society.
2. Individually available latent options. This can be understood as the subset of latent options that is actually available to a particular person. This availability is based, for example, on the availability of cultural capital (Bourdieu, 2016) or physical constitution.
3. Manifest options. If individually available latent options are actually exercised, manifest options arise.
4. Potential options. These form a residual category of options that go beyond latent options. These are options that can be found in other societies or may exist in the future. They can therefore also be referred to as contingent options.

When Dahrendorf talks about the expansion of options as the basis for a society with growing life chances, this means in particular converting potential options into latent options, as well as latent options into individually available latent options. However, the conversion of individually available latent options into manifest options is beyond the scope of political responsibility, because whether a person makes use of a latent option available to them is their own responsibility.

4.3 Summary

This chapter has made it clear that a normative orientation toward life chances does not attempt to resolve the tension between the individual pursuit of options and social integration through ligatures, nor does it favor the unrestricted prioritization of the individual or social determinism. Rather, it attempts to turn this—thoroughly conflictual—tension into something productive. The concept of life chances—as a space of possibility for individual development in the context of social structures—enables a middle ground: on the one hand, options require ligatures to become meaningful. On the other hand, ligatures require options in order not to end up in normative rigidity. Freedom thus does not mean the absence of ties, but rather a reflective shaping of spaces for action.

The differentiation of ligatures—for example, into moral, ethical, explicit, or physically mediated forms—as well as the structuring of options into (individually available) latent, manifest, and potential forms represents an analytical tool that allows us to deal in a differentiated manner with the conditions for successful self-realization (or its non-realization) under complex social conditions. The concept of life chances can also be understood as a plea for a socially informed ethics: it pursues the goal of designing structures in such a way that as many options as possible are created that can be associated with meaning. This meaning arises from the reference to options to voluntarily enter into ethical ligatures, especially those with an internal orientation. In the following chapter, we will take up this idea and ask how this perspective can be related to a (neo)pragmatic worldview, which in turn provides the basic framework for thinking about impulses (in the sense of new metaphors) for the vocabulary of sustainable development.

References

Ackermann, U. (2020). *Das Schweigen der Mitte: Wege aus der Polarisierungsfalle.* WBG.

Aristotle. (2009). Politik. Neuausgabe. Reinbek bei Hamburg: Rowohlt (Nach der Übersetzung von Franz Susemihl mit Einleitung, Bibliographie und zusätzlichen Anmerkungen von Wolfgang Kullmann).

Berger, P. L. (2017 [1963]). *Einladung zur Soziologie: Eine humanistische Perspektive* (2., ergänzte Auflage). UVK; UVK/Lucius. (Original work published 1963)

Berr, K., Jenal, C., Koegst, L., & Kühne, O. (2022). *Noch mehr Sand im Getriebe? Kommunikations- und Interaktionsprozesse zwischen Landes- und Regionalplanung, Politik und Unternehmen der Gesteinsindustrie.* Springer VS.

Bourdieu, P. (2016). *La distinction: Critique sociale du jugement. Le Sens commun.* Editions de Minuit.

Bovone, L. (1982). Libertà e utopia in Marcuse e Dahrendorf. *Studi di Sociologia, 20*(3/4), 273–296.

Dahrendorf, R. (1957). *Soziale Klassen und Klassenkonflikt in der industriellen Gesellschaft.* Enke.

Dahrendorf, R. (1968). *Pfade aus Utopia: Arbeiten zur Theorie und Methode der Soziologie.* Piper.

Dahrendorf, R. (1972). *Konflikt und Freiheit: Auf dem Weg zur Dienstklassengesellschaft.* Piper.

Dahrendorf, R. (1979). *Lebenschancen: Anläufe zur sozialen und politischen Theorie. Suhrkamp-Taschenbuch: Vol. 559.* Suhrkamp.

Dahrendorf, R. (1983). *Die Chancen der Krise: Über die Zukunft des Liberalismus.* Deutsche Verlags-Anstalt DVA.

Dahrendorf, R. (1992). *Der moderne soziale Konflikt: Essay zur Politik der Freiheit.* Deutsche Verlags-Anstalt DVA.

Dahrendorf, R. (1994). Das Zerbrechen der Ligaturen und die Utopie der Weltbürgergesellschaft. In U. Beck & E. Beck-Gernsheim (Eds.), *Riskante Freiheiten: Individualisierung in modernen Gesellschaften* (pp. 421–436). Suhrkamp.

Dahrendorf, R. (2006). *Versuchungen der Unfreiheit: Die Intellektuellen in Zeiten der Prüfung.* C.H. Beck.

Dahrendorf, R. (2007). *Auf der Suche nach einer neuen Ordnung: Vorlesungen zur Politik der Freiheit im 21. Jahrhundert* (4th ed.). *Krupp-Vorlesungen zu Politik und Geschichte am Kulturwissenschaftlichen Institut im Wissenschaftszentrum Nordrhein-Westfalen: Vol. 3.* C.H. Beck.

Hadot, P. (2005). *Philosophie als Lebensform: Antike und moderne Exerzitien der Weisheit* (2., ungekürzte deutsche Neuauflage). Fischer.

Horn, C. (2014). *Antike Lebenskunst: Glück und Moral von Sokrates bis zu den Neuplatonikern* (3rd ed.). *Beck'sche Reihe: Vol. 1271.* C.H. Beck.

Kamlah, W. (1973). *Philosophische Anthropologie: Sprachkritische Grundlegung und Ethik. BI-Hochschultaschenbücher: Vol. 238.* Bibliographisches Institut.

Kant, I. (1983 [1784]). Idee zu einer allgemeinen Geschichte in weltbürgerlicher Absicht. In W. Weischedel (Ed.), *Kant. Werke. Band 9* (pp. 31–50). WBG.

Kocka, J. (2004). Dahrendorf in Perspektive. *Soziologische Revue, 27*(2), 151–158.

Krämer, H. (1992). *Integrative Ethik*. Suhrkamp.

Krämer, H. (1998). Integrative Ethik. In J. Schummer (Ed.), *Glück und Ethik* (pp. 93–107). Königshausen & Neumann.

Kühne, O. (2019). Die Produktivität von Landschaftskonflikten – Möglichkeiten und Grenzen auf Grundlage der Konflikttheorie Ralf Dahrendorfs. In K. Berr & C. Jenal (Eds.), *Landschaftskonflikte* (pp. 37–49). Springer VS.

Kühne, O. (2024a). Neopragmatische Annäherungen an horizontale Geographien – zum Stand der Dinge, zu Potenzialen und Herausforderungen. In F. Weber, O. Kühne, & J. Dittel (Eds.), *Transformation Processes in Europe and Beyond: Perspectives for Horizontal Geographies* (pp. 23–48). Springer VS.

Kühne, O. (2024b). *Redescribing Horizontal Geographies: A Neopragmatist Approach to Spatial Contingency, Complexity, and Relationships*. Springer International.

Kühne, O. (2025). *Schöner Scheitern in der Wissenschaft: Karikaturen von der Hinterbühne des Betriebs*. Springer.

Kühne, O., Berr, K., Edler, D., Lohmann, P., & Schuster, K. (2025). *Neopragmatismus – inverse Landschaft – (karto)graphische Darstellung: Von der Kritik zur Weiterentwicklung*. Springer VS.

Kühne, O., Berr, K., & Jenal, C. (2022). *Die geschlossene Gesellschaft und ihre Ligaturen: Eine Kritik am Beispiel 'Landschaft'*. Springer VS.

Kühne, O., Berr, K., & Koegst, L. (2023). Contingency and Landscape: Basic Considerations on Graphic and Cartographic Representations in Recourse to the Concept of Inverse Landscapes as a Contribution to Deviant Cartographies with Examples on Louisiana. *KN—Journal of Cartography and Geographic Information, 73*(3), 225–236. https://doi.org/10.1007/s42489-023-00145-7

Kühne, O., Berr, K., Schuster, K., & Jenal, C. (2021). *Freiheit und Landschaft: Auf der Suche nach Lebenschancen mit Ralf Dahrendorf*. Springer.

Kühne, O., & Koegst, L. (2023). *Land Loss in Louisiana: A Neopragmatic Redescription*. Springer VS.

Kühne, O., & Leonardi, L. (2020). *Ralf Dahrendorf: Between Social Theory and Political Practice*. Palgrave Macmillan. https://doi.org/10.1007/978-3-030-44297-2

Kühne, O., Schönwald, A., & Jenal, C. (2022b). Bottom-up Memorial Landscapes Between Social Protest and Top-Down Tourist Destination: The Case of Chicano Park in San Diego (California)—An Analysis Based on Ralf Dahrendorf's Conflict Theory. *Landscape Research, 48*(5), 615–631. https://doi.org/10.1080/01426397.2022.2069731

Kühne, O., Silina, I., & Berr, K. (2024). *Nachhaltige Entwicklungen und Lebenschancen*. Springer Fachmedien Wiesbaden. https://doi.org/10.1007/978-3-658-45998-7

Lohmann, P., Kühne, O., & Berr, K. (2024). *Landschaft und Gefühl – eine neopragmatistische Redeskription auf Grundlage der Philosophie Fichtes: Ein Beitrag zur humanistischen Geographie*. SpringerVS.

Luhmann, N. (1993). Die Moral des Risikos und das Risiko der Moral. In G. Bechmann (Ed.), *Risiko und Gesellschaft: Grundlagen und Ergebnisse interdisziplinärer Risikoforschung* (pp. 327–338). Westdeutscher Verlag.

Luhmann, N. (2016). *Die Moral der Gesellschaft* (D. Horster, Ed.) (4th ed.). Suhrkamp.

Mackert, J. (2010). Opportunitätsstrukturen und Lebenschancen. *Berliner Journal Für Soziologie, 20*(3), 401–420. https://doi.org/10.1007/s11609-010-0135-7

Niedenzu, H.-J. (2001). Konflikttheorie: Ralf Dahrendorf. In J. Morel, E. Bauer, T. Maleghy, H.-J. Niedenzu, M. Preglau, & H. Staubmann (Eds.), *Soziologische Theorie: Abriß ihrer Hauptvertreter* (7th ed., pp. 171–189). R. Oldenbourg Verlag.

Riesman, D. (1950). *The Lonely Crowd*. Yale University Press.

Thiel, A., & Kühne, O. (2024). *Theorien sozialer Konflikte – vom Überblick zur Weiterentwicklung*. Springer VS.

Weber, M. (1976 [1922]). *Wirtschaft und Gesellschaft. Grundriß der verstehenden Soziologie*. Mohr Siebeck.

Chapter 5
From a Summary of the Theoretical Background to Criteria for Further Developing the Vocabulary of Sustainable Development

In Chaps. 2–4, we have established a theoretical framework from different perspectives with which we will address the complexity of normative negotiation processes in the context of sustainable development. In Chap. 2, we introduced two analytical tools, the three-world theory and the typology of scientific criticism, which help us to reflect on the complex relationships between the individual, society, and materiality, as well as the relationships between knowledge, norms, and practice. Chapter 3 introduced neopragmatism—particularly in line with Richard Rorty (especially Rorty, 1979, 1989, 2023)—as a metatheoretical frame of reference in which theory is no longer understood as a reflection of the world but as cultural practice. This approach understands language, experience, and social order in a dynamic interrelationship. In Chap. 4, we introduced the concept of life chances as an analytical approach to the relationship between the individual and society, while also sharpening the normative perspective: individual freedom and social bonds are not understood here as opposites, but as relational, reciprocal conditions that can be shaped in ways that are both individually beneficial and socially productive. What these three chapters have in common is the concern to overcome reductionist models, for example, in the form of epistemic objectification, political totalization, or moral (self-)aggrandizement. This is replaced by an understanding of the world that is not characterized by the 'discovery' or 'establishment' of order, but by orientation; it is not 'truth' that is to be found and then used as justification for 'unavoidable' measures (although the direction of causality can also be reversed), but rather it is a matter of proving oneself in conversation and solving problems.

Popper's three-world theory (Popper, 1979; Popper & Eccles, 1977) allows us, as a quasi-ontology, to understand descriptions of the world of cultural and social content (World 3), individual experience (World 2), and material foundations (World 1) as distinct but (always via World 2) interacting levels. With Davidson's concepts of triangulation on the one hand and the principle of charity on the other, it becomes clear how much understanding is always also an act of mutual interpretation. This is

O. Kühne et al., *Sustainable Development and Life Chances*, Neopragmatic Horizontal Geographies 3, https://doi.org/10.1007/978-3-032-20182-9_5

never complete and can always be revised. In this way, an understanding of communication emerges that is characterized both by the initial acceptance of difference and by the desire for connectivity. Such a perspective can prove useful for the principle of sustainable development when it comes to not exhausting oneself in global target systems but rather remaining open to local and regional, subjective, and culturally pluralistic approaches. It also remains open to the constant revisability of one's own positions—through new metaphors and openness to redescriptions. At this point, the relevance of Rorty's vocabulary theory becomes clear: vocabularies are expressions of historically grown, partially shared, and always contingent approaches to the world. Vocabularies not only structure what can be considered a problem, they also structure what can be developed as a solution. If both lose their usefulness, the vocabulary can be redescribed. Redescription thus becomes the central mode of theoretical and practical renewal. Redescription does not mean replacing a 'false' description with a 'true' one. Redescription means creating a contingent interpretation of the world that promises greater usefulness than vocabularies that have become 'time-honored'. But even redescribed vocabularies must prove their usefulness and are always subject to redescribing themselves. Usefulness in this sense is not measured by an objective and universally valid standard. Rather, it is measured by the ability to enable understanding, action, and the avoidance of suffering—in short, to enlarge the "we." In the context of sustainable development, where technical knowledge, political goal-setting, individual lifestyles, private self-creation, and public solidarity are intertwined, such a pragmatic attitude promises to offer a useful alternative—in the above sense—to dogmatic, top-down-oriented conceptual politics.

At this point, the potential of the theory of life chances (Dahrendorf, 1979)—especially based on Dahrendorf—becomes clear: options and ligatures do not appear as antipodes between which people and society must choose. Rather, they form complementary poles of successful living that are constitutively linked to each other. Ligatures, understood as deep social bonds, have an ambivalent effect: on the one hand, they provide orientation and stability, but on the other hand, they can also become authoritarian and prevent people from creatively shaping their own lives. Options, in turn, outline the scope of action available to individuals. They become effective when they resonate with meaning, i.e., ligatures. Life chances thus arise from the productive intertwining of ligatures and options: options remain formal until they are (voluntarily) linked to ligatures, whereby they become substantial options. This is an idea that can be intertwined with Rorty's concept of the liberal ironist, because she too lives in a state of tension, namely that of private self-creation in the awareness of the contingency of all her convictions, as well as of language, self, and society, and of public solidarity as an attempt to avoid pain and expand the "we." This can be intertwined with options and ligatures: the liberal ironist strives to expand her individually latent options, voluntarily entering into ligatures that ethically enable her to act responsibly and in solidarity. The direction of the voluntarily entered into ligatures is inward and is explained explicitly when necessary. She acts toward others based on moral ligatures of compassion, not resentment, because she is aware that there is no definitive vocabulary—not even one that she represents. In this respect, she is also unfamiliar with outward-looking moral ligatures. Pain is not only pain caused by exclusion or being rewritten, but also physical pain—in this respect, she is aware of the physical vulnerability of human beings. This means that in a vocabulary of sustainable development that seeks to do justice to this figure of thought of the liberal ironist's expansion of the resonances of options and ligatures, it is important to recognize individual needs for self-creation as well as collective goals of environmental justice as mutually legitimate perspectives. This should not be done through abstraction and reference to universal principles, but through practical negotiation in the light of concrete problems that are recognized as such by the community.

In such an intertwining, the potential of a neo-pragmatic approach becomes clear: it undermines simple claims to truth and moral rigorism. In doing so, it does not tip over into arbitrariness but rather formulates a framework for assessment based on the avoidance of pain and the demand for inclusion and tolerance. Theory is understood here as a practice of conversation—not as expert discourse, but as a continuous process of understanding between vocabularies. In this sense, sustainability is understood not as a state but as a process. It can be understood as a never-ending, conflictual, but potentially solidarity-based struggle for inclusive futures.

Based on this understanding, aspects can be formulated that could guide the further development of the vocabulary of sustainable development:

1. A departure from representationalism: Sustainable development is not understood as a quasi-inevitable derivation of a supposedly objective image of a given world, but rather as an error-prone tool in the context of current challenges.
2. Triangulation of worlds: Sustainable development is understood as relating to the problems in World 1 (matter), World 2 (experience), and World 3 (culture) and avoids teleological exaggerations or utopian totalizations.
3. Constructive criticism: Critical debate does not remain stuck in negation or exposure but rather aims to open up new spaces for discussion and enable new solutions to problems; in short, to produce new metaphors.
4. Avoiding pain and humiliation: Redescriptions are not intended to destroy or marginalize existing self-relationships and other vocabularies. Neopragmatic redescriptions transform gently, inclusively, empathetically, and respectfully.
5. Primacy of the new metaphor over theory: A metaphorical extension allows for cautious transitions and facilitates cultural resonance with less pain. In this respect, new metaphors often form a less painful alternative to redescription.
6. Benefit orientation: Redescriptions in particular must be characterized by their problem-solving ability. This is not to be understood in terms of technocratic efficiency but rather against the backdrop of a shared way of life.
7. Diversity instead of truth: Progress does not come about through agreement on the objectively 'best' description or the only 'most useful' measure. It comes about through the pluralization of discourse in freedom. This also means not having to agree.
8. Mutual reference between self-creation and solidarity: Sustainable development must enable individual self-creation and at the same time promote community

ties. This does not mean a balancing act but rather forms a structural coordinate system for the creation of meaning in the resonance between options and voluntarily chosen ligatures.

9. Sensitivity to conflict theory: Conflicts in the process of sustainable development are inevitable; they can be a functional struggle for suitable solutions if they are not carried out in the field of moral destruction of resentment. Conflicts can be understood as an expression of competing vocabularies that remain in dialogue with each other—and mutually recognize the legitimacy of the other vocabulary.
10. Integration and reusability: New vocabularies can be incorporated into future redescriptions. Every new metaphor, every new vocabulary, means not seeing oneself as the end point but as a finite contribution to the conversation of the many.
11. Appreciation of locally embedded practices: Sustainable development is not understood as universally boundless; it must take its limitations into account in its vocabulary—also in a spatial sense. This means that it also takes regionally differentiated interpretations of the world seriously.
12. Procedural and equal opportunity: Fair procedures and the expansion of the field of individually available options contribute to sustainable development.

We will use these 12 criteria to examine the SDGs in the following.

References

Dahrendorf, R. (1979). *Lebenschancen: Anläufe zur sozialen und politischen Theorie. Suhrkamp-Taschenbuch: Vol. 559*. Suhrkamp.

Popper, K. R. (1979). Three Worlds: Tanner Lecture, Michigan, April 7, 1978. *Michigan Quarterly Review, 18*(1), 141–167.

Popper, K. R., & Eccles, J. C. (1977). *The Self and Its Brain: An Argument for Interactionism*. Springer.

Rorty, R. (1979). *Philosophy and the Mirror of Nature*. Princeton University Press.

Rorty, R. (1989). *Contingency, Irony, and Solidarity*. Cambridge University Press.

Rorty, R. (2023). *Pragmatismus als Antiautoritarismus* (J. Schulte, Trans.) (E. Mendieta, Ed.). Suhrkamp.

Chapter 6
Historical Review of Sustainability, the SDGs, and Differentiated Approaches to Sustainable Development

The Sustainable Development Goals (SDGs) are at the heart of the United Nations' 2030 Agenda, which was adopted in 2015. Since then, this agenda has symbolized the call to "transform our world." All countries were called upon to pursue the 17 sustainability goals and integrate them into their national development strategies. The emergence of the SDGs was a lengthy process—from the development of the idea to the recognition of the need for action to implementation at the international and national levels.

Although the term 'sustainability' has only been around for a few centuries, the basic ideas of sustainable development can already be found in pre-modern cultures. In ancient agricultural societies in particular, the conscious use of nature and resources was considered vital (Spindler, 2012). However, environmental awareness was not seen everywhere as essential for survival. As long as people lived as nomads, they were able to avoid overuse by moving on when resources became scarce (Rogall, 2008). However, with the advent of settled life and agriculture, pressure on natural resources such as soil, forests, and water increased. Nevertheless, a romanticized view of pre-industrial societies should be viewed critically. These societies also caused environmental damage or only avoided it because they lacked the technical means to exploit resources (Diamond, 2011; Küster, 2013 [1995]; Radkau, 2000; Sieferle, 1984). It was not until the industrial age that profound, often irreversible environmental damage occurred. Technical progress, increasing mechanization, the increased use of chemicals in agriculture, and global trade networks had a major ecological impact (Förstner & Köster, 2018). These were not only the consequences of a consumption-oriented lifestyle in the Global North but also of a growing world population (Schenk, 2011). At the same time, a strict juxtaposition of 'nature' and 'industry' is not very productive when it comes to sustainable development. Industry can also have positive effects on the environment. For example, sustainable forestry was only possible thanks to industrial progress, as coal covered energy needs and

---

This chapter essentially summarizes our remarks in Kühne et al. (2024).

O. Kühne et al., *Sustainable Development and Life Chances*, Neopragmatic Horizontal Geographies 3, https://doi.org/10.1007/978-3-032-20182-9_6

thus reduced excessive wood consumption (Küster, 2005, p. 146). Fertilizers made from industrial waste such as potash salt or Thomas meal also improved the food situation, while the railroad enabled nationwide transport (Küster, 2005, p. 148). In addition, industrialization gave rise to important environmental technologies such as sewage treatment plants, waste disposal, and air filtration systems, for which the Ruhr region serves as an example Klicken oder tippen Sie hier, um Text einzugeben. Finally, there are many examples of incorrect predictions and environmental alarms, such as the debate about "forest dieback" in the 1970s (Gethmann, 2009; Küster, 2005; Radkau, 2017).

6.1 History of Sustainable Development

Forestry is one of the most important origins of the concept of sustainability. The term "sustain" was first used in this field in the eighteenth century. At that time, the principle of sustainability established itself as a central pattern of action in forestry (von Hauff, 2021). The first known written evidence of the principle of sustainability comes from Hans Carl von Carlowitz (1645–1714), Chief Mining Administrator in Freiberg, Saxony.

In his 1713 work *Sylvicultura oeconomica*, Carlowitz emphasized the need to use wood reserves with foresight. At that time, Europe's forests had already been severely damaged by massive overexploitation, and wood shortages were becoming an increasingly serious problem (Radkau, 1983). Carlowitz therefore demanded that only as much wood be felled as could be regrown through reforestation. He distinguished between untouched forests and artificially managed areas, where targeted human intervention was intended to shape the forest in such a way that it would provide sufficient resources in the long term. Sustainable forest management was to be the basic principle of this targeted use and respect for the natural cycle. The goal was to ensure the stability of the forest ecosystem, with Carlowitz explicitly calling for 'sustainable use of forests'. He advocated for continuous timber use that is designed for the long term: "to cultivate the timber so that there is continuous, consistent, and sustainable use" (Carlowitz, 2022 [1732], p. 17). In doing so, he placed the preservation of resources while simultaneously utilizing them economically at the center of his considerations.

The principle of sustainability, which originated in forestry, was later applied to numerous other environmental problems and developed into a guiding principle in politics, economics, and society (Spindler, 2012). Towards the end of the nineteenth century, the German model of sustainable forestry was translated into English and became known as 'sustained yield forestry'. This gave *Sylvicultura oeconomica* historical significance, as Carlowitz was the first to link the concept of 'sustainability' with the idea of providing for future generations (Grober, 2013).

To understand today's sustainability discourse, it is worth taking a look at the environmental debates of the 1960s and 1970s. During this period, a new awareness of the limited nature of environmental resources emerged. Particularly influential was the discourse on the 'limits to growth', which emphasized the connection between the environment and social development. At that time, there were calls to limit and stabilize economic and population growth—a demand that already anticipated aspects of today's idea of sustainability, even though the term 'sustainable development' was not yet in use (Weiland, 2007). Since the 1960s, scientists have been warning about growing environmental problems and their far-reaching consequences. Since then, they have been calling for greater consideration to be given to the environmental costs of economic growth (von Hauff, 2021). An important impetus came in 1962 with the book '*Silent Spring*' by biologist Rachel Carson, which is considered to have triggered the global environmental movement. In it, she vividly describes the ecological damage caused by pesticides and herbicides in agriculture, which severely disrupt the ecological balance that has developed over millions of years. She focuses in particular on the mass death of birds due to chemical poisoning (Carson, 1962). The book caused an international sensation and led to intense debates about man-made environmental destruction. During this period, the idea of 'limits to growth' also gained increasing importance and was discussed as a counter-model to the dominant growth paradigm (Weiland, 2007).

Another work that highlighted the downsides of economic growth early on was '*The Costs of Economic Growth*' by Edward J. Mishan. In it, the British economist emphasizes that economic growth often leads to environmental destruction, such as air and water pollution, deforestation, and the depletion of natural resources (Mishan, 1967). In his view, traditional economic analyses often ignore or downplay these follow-up costs. Mishan also emphasizes that economic growth can exacerbate social inequalities and destroy traditional ways of life. He calls for greater consideration of ecological and social impacts in economic decision-making processes. He also considers the health impacts of environmental pollution and the negative social effects of urbanization and industrialization to be among the downsides of unbridled growth. Overall, Mishan questions the widespread assumption that economic growth automatically leads to a better quality of life. He advocates a more comprehensive view that considers all the costs of growth, including the impact on the environment, health, and social structures (Mishan, 1967).

A particularly significant contribution to the sustainability debate was the 1972 report '*The Limits to Growth*' by the Club of Rome. Dennis Meadows and his team issued an urgent warning about the consequences of unchecked economic and demographic growth. They saw this as a key cause of future food shortages, the exploitation of natural resources, and increasing environmental pollution (Meadows et al., 1972). Using a complex mathematical model, they examined global development trends and their interactions over a period of 100 years (Schwartz, 2016). Their central warning was that if the existing growth rates of population, industrialization, environmental pollution, food production, and raw material consumption continue unabated, the planetary limits to growth will be exceeded in the coming century at the latest. This could lead to a rapid and irreversible decline in population and industrial production. Nevertheless, the authors emphasized that a reversal was possible to create an ecological and economic balance that could be sustained in the long term. The prerequisite for this was a global system that would ensure the preservation of natural resources while keeping individual development opportunities open for all people (Meadows et al., 1972). The report aimed to bring about a rethinking of international policy. It quickly became a catalyst for the global sustainability discourse and attracted considerable attention at the first United Nations Conference on the Human Environment in Stockholm that same year.

One of the most important milestones in the development of the concept of sustainability is the *Brundtland Report, 'Our Common Future'*, published by the United Nations World Commission on Environment and Development. This report brought the issue of sustainability to the forefront of international politics for the first time (von Hauff, 2021). Against the backdrop of increasing environmental problems and stagnating development in the Global South, the commission was tasked with developing strategies and recommendations for future-oriented policies. The final report presented a concept for sustainable development that combines environmental and development goals (Weiland, 2007). The report defined sustainable development as "development that meets the needs of the present generations without compromising the ability of future generations to meet their own needs" (World Commission on Environment and Development [WCED], 1987, p. 14). The focus is on the needs of the present generation and responsibility towards future generations (von Hauff & Kleine, 2009). The Brundtland Report spread an optimistic view of the possibilities of sustainable growth. It regarded technological progress as both a driver of economic growth and a means of protecting the environment: "What is needed now is a new era of economic growth—growth that is forceful and at the same time socially and environmentally sustainable" (WCED, 1987, p. 14). Despite worldwide recognition, the report remained vague on many points, leaving room for different interpretations. Nevertheless, it sparked a broad international discussion on the practical implementation of sustainable development. In addition, the Brundtland Commission recommended a world conference, which took place in 1992 and is considered the starting point of the so-called 'Rio Process' (von Hauff, 2021).

The *UN Conference on Environment and Development* in Rio de Janeiro, initiated by the Brundtland Commission in 1992, contributed significantly to the global spread of the concept of sustainability (Spindler, 2012). At the conference, 178 countries adopted key documents (Michelsen, 2012):

- Rio Declaration on Environment and Development, which emphasizes the right to development for present and future generations,
- the Framework Convention on Climate Change to reduce greenhouse gas emissions,
- Convention on Biological Diversity to preserve biological diversity,
- Forest Declaration with principles for sustainable forest management,
- Agenda 21 as a global action program for environmental and development policy.

Agenda 21 in particular pursued the goal of promoting sustainable development worldwide at the local and global levels (UNCED, 1992). However, since no binding commitments were agreed upon, implementation was often slow. Criticism was also directed at the strong market orientation and the non-binding nature of many of the objectives. Numerous compromises between different national interests contributed to the vague wording of the resolutions (Michelsen, 2012). Following the Rio Conference, further international meetings were organized, including the 1994 World Population Conference, the 1995 World Social Summit, and the 1997 Climate Conference (von Hauff, 2021). A significant outcome of the Rio Conference was the establishment of the UN Commission on Sustainable Development (CSD), which has since monitored progress in implementing sustainable development (Michelsen, 2012).

In 2000, the United Nations adopted the *Millennium Declaration*, which led to the formulation of the eight *Millennium Development Goals (MDGs)* in 2001. These goals focused on poverty reduction and social aspects (Köhler, 2015). These included poverty reduction, education for all, gender equality, reducing child mortality, improving maternal health, combating disease, and environmental and development partnerships. The assessment of the MDGs was mixed: although extreme poverty declined in many regions, other goals were not achieved (United Nations, 2015). Criticism was leveled at the uneven distribution of progress and inadequate indicators for measuring success (German Society for the United Nations, n.d.). Nevertheless, the MDGs strengthened international cooperation and were incorporated into the development of the successor agenda: Agenda 2030 with the Sustainable Development Goals (SDGs) (Köhler, 2015).

The Rio Process was continued in subsequent years through further UN conferences. The focus was on the *2002 World Summit on Sustainable Development in Johannesburg*, which was intended to assess the progress made since the 1992 conference (Spindler, 2012). The summit adopted the Johannesburg Declaration and an action plan focusing on poverty reduction, environmental and resource protection, and the promotion of global partnerships. In addition, the importance of education as a key factor for sustainable development was emphasized. This led to the proclamation of the World Decade of Education for Sustainable Development by the United Nations, coordinated by UNESCO (Michelsen, 2012). However, critics complained about the lack of binding commitments and timetables. The weak participation of some industrialized countries and the lack of effective enforcement mechanisms were also criticized. In addition, it was emphasized that it had not been possible to legally anchor social and ecological interests at the global level and to create a counterweight to institutions such as the World Trade Organization (WTO) (Unmüßig, 2003, p. 15).

The *UN Rio + 20 Conference in 2012* followed on from the earlier summits. The aim was to evaluate the progress made in the 20 years since the first Rio Conference and to renew efforts towards sustainable development. Under the motto 'Green Economy', a call was made for an economy based on ecological sustainability and poverty reduction (KOM, 2011). Numerous new initiatives were launched, including more than 700 private-public partnerships under the motto "Sustainable Development in Action" (Brühl & Rosert, 2014). The conference decided to develop universal, albeit legally non-binding, sustainability goals by 2015. This resulted in Agenda 2030 with its 17 global Sustainable Development Goals (SDGs) (von Hauff, 2021). These are intended to strengthen the global promotion of sustainable development and contribute to a certain degree of international standardization.

6.2 Sustainable Development Goals

With the adoption of Agenda 2030 in September 2015 at the UN Summit in New York under the title '*Transforming our world: The 2030 Agenda for Sustainable Development*', the international community committed itself to a comprehensive program of action dedicated to people, the planet, and prosperity (United Nations, 2015). This agenda combines two previously separate UN processes: the so-called Rio Process, which dates back to Agenda 21 of 1992, and the implementation process of the Millennium Development Goals. In terms of sustainable development, Agenda 2030 is considered a significant milestone in the recent history of the United Nations (von Hauff, 2021). It goes far beyond the Millennium Development Goals (MDGs) in that it considers not only social aspects but all three dimensions of sustainability—economic, social, and environmental—as well as peace and international partnership (Martens, 2018). Unlike the MDGs, the Sustainable Development Goals (SDGs) are not aimed exclusively at developing countries but apply to all countries—regardless of whether they are classified as developing, emerging, or industrialized countries (United Nations, 2015).

Agenda 2030 comprises 17 global goals, which are specified in a total of 169 sub-goals. These are to be implemented by 2030 through appropriate political measures. The sustainability concept of the agenda is based on three equal dimensions: a productive global economy that utilizes technical progress without neglecting environmental and social aspects (Wölkner, 2019). The complexity of the agenda is particularly evident in the close interconnection between the individual SDGs, which not only overlap within the three dimensions of sustainability but also in the five overarching guiding principles, the so-called 'core messages' (United Nations, 2015):

- People: End poverty and hunger, and ensure dignity, equality, and a healthy environment for all.
- Planet: Protecting natural resources and ecosystems for present and future generations.
- Prosperity: Promoting prosperity through sustainable economic and technological progress harmonizing with nature.
- Peace: Building peaceful, just, and inclusive societies that are free from violence and fear.
- Partnership: Strengthening global partnerships and cooperation to achieve the goals.

In addition, Agenda 2030 sets out five key principles for implementation (United Nations, 2015):

- Universality: The agenda applies equally to all countries.
- Indivisibility: The goals must be implemented as a whole, not selectively.
- Leave no one behind: all people should benefit from progress, especially the most disadvantaged.
- Accountability: Progress toward achieving the goals must be regularly reviewed and made transparent, both nationally and internationally.

- Partnership: All social actors—governments, businesses, cities and municipalities, civil society, and academia—share responsibility for implementation.

For an overview of the SDGs (United Nations, 2015), see Box 6.1.

Box 6.1: The Sustainable Development Goals (SDGs)

SDG 1: End poverty in all its forms everywhere.
SDG 2: End hunger, achieve food security and improved nutrition and promote sustainable agriculture.
SDG 3: Ensure healthy lives and promote well-being for all at all ages.
SDG 4: Ensure inclusive and equitable quality education and promote lifelong learning opportunities for all.
SDG 5: Achieve gender equality and empower all women and girls.
SDG 6: Ensure availability and sustainable management of water and sanitation for all.
SDG 7: Ensure access to affordable, reliable, sustainable and modern energy for all.
SDG 8: Promote sustained, inclusive, and sustainable economic growth, full and productive employment and decent work for all.
SDG 9: Build resilient infrastructure, promote inclusive and sustainable industrialization and foster innovation.
SDG 10: Reduce inequality within and among countries.
SDG 11: Make cities and human settlements inclusive, safe, resilient and sustainable.
SDG 12: Ensure sustainable consumption and production patterns.
SDG 13: Take urgent action to combat climate change and its impacts.
SDG 14: Conserve and sustainably use the oceans, sea, and marine resources for sustainable development.
SDG 15: Protect, restore and promote sustainable use of terrestrial ecosystems, sustainably manage forests, combat desertification, halt and reverse land degradation, and halt biodiversity loss.
SDG 16: Promote peaceful and inclusive societies for sustainable development, provide access to justice for all and build effective, accountable and inclusive institutions at all levels.
SDG 17: Strengthen the means of implementation and revitalize the Global Partnership for Sustainable Development.

Some of the SDGs refer directly to economic issues, such as SDG 8 (Decent work and economic growth), SDG 9 (Industry, innovation and infrastructure), and SDG 12 (Responsible consumption and production). Other goals address social aspects, including SDG 1 (No Poverty), SDG 5 (Gender Equality), and SDG 4 (Quality Education). In addition, there are goals that focus on protecting the environment, such as SDG 13 (Climate Action), SDG 14 (Life Below Water), and SDG 15 (Life on Land).

There is broad agreement that there are complex interactions between the SDGs and their sub-goals. However, these can only be exploited positively if all goals are pursued jointly—especially in the area of sub-goals (von Hauff, 2021). Conflicts of interest are related to traditional development paths that focus on economic growth, often at the expense of the environment. Developed countries in particular have a higher average level of human prosperity, but this is associated with high resource consumption and environmental pollution. To achieve SDG 12 (Responsible Consumption and Production), this consumption must be significantly reduced (Pradhan et al., 2017).

In the past, economic growth and an increase in the Human Development Index (HDI) have led to improvements in health and nutrition, but at the same time to an increase in greenhouse gas emissions and food waste (Pradhan et al., 2013). Therefore, there is a growing focus on the need for social transformation to break these close links and enable the successful implementation of the SDGs (Pradhan et al., 2017).

The 17 main goals and 169 sub-goals of the 2030 Agenda ultimately represent a political compromise. Nevertheless, they provide important guidance for global policy that does not focus on nationalist isolationism but rather shapes globalization in the sense of sustainable prosperity for all people (Michelsen, 2017).

6.3 Differentiated Approaches to Understanding Sustainability

The discussion on sustainability has become much more differentiated since the publication of the Brundtland Report in 1987 ('Our Common Future'). Various interpretations and operationalizations have been developed, some of which differ considerably from one another (Ekart, 2005; Grunwald & Kopfmüller, 2012; Pufé, 2017; von Hauff & Kleine, 2009). Before defining our own understanding of sustainability in more detail, it therefore seems necessary to outline the central lines of discourse:

- Dimensions of sustainable development: The most widely used model for illustrating sustainable development is the three-pillar model, which encompasses the areas of ecology, economy, and social issues. The goal is to achieve a balance between environmental conservation, social justice, and economic performance, with each dimension having as little impact on the others as possible.

  However, there are alternative approaches. Some theories emphasize a monodimensional view in which the environment is given priority because it forms the indispensable basis of human life. Other concepts, on the other hand, expand the dimensions of sustainability to include politics (keyword: good governance) or culture, with the latter being understood as a driver for sustainable action (Kühne et al., 2024).
- Substantive versus procedural sustainability: Two fundamental perspectives shape the sustainability discourse: the substantive and the procedural understanding. The substantive understanding follows a normative approach to government control. Here, sustainability goals are set, from which concrete strategies and measures are derived (top-down).

This contrasts with the procedural understanding, which relies on an open search process. Goal definitions and options for action should be flexibly adaptable within the framework of social self-organization. Sustainability is understood less as a fixed state and more as a dynamic process (Kühne et al., 2024).

- Strategies for sustainable development: efficiency, consistency, and sufficiency (von Hauff & Kleine, 2009).

  – Efficiency strategy: The goal is to minimize the use of materials and energy per product unit, for example, through energy-efficient production processes.
  – Consistency strategy: The focus here is on material substitution, i.e., replacing fossil raw materials with renewable alternatives or low-emission materials.
  – Sufficiency strategy: This strategy focuses on limiting consumption and production to reduce ecological and social side effects. It questions whether certain consumer goods are necessary at all.

    In practice, a combination of these approaches is often sought, with sufficiency strategies often being the most difficult to implement socially.

- Weak versus strong sustainability: Another key point of discussion concerns the relationship between natural and human capital. Proponents of strong sustainability call for the conservation of natural resources within their regeneration limits. In contrast, the concept of weak sustainability accepts an exchange between natural capital and man-made capital, as long as the overall benefit for future generations is preserved (Ekart, 2005; Grunwald & Kopfmüller, 2012; Pufé, 2017; Rogers et al., 2012; von Hauff & Kleine, 2009).

  Hybrid approaches are also being discussed between these two poles, for example, with graduated substitution options depending on the type of resource.
- Discourses on justice in the context of sustainability: Sustainability is closely linked to questions of justice, with very different concepts of justice feeding into the debate (Ekart, 2005; Grunwald & Kopfmüller, 2012; Pufé, 2017; Rogers et al., 2012; von Hauff & Kleine, 2009). These can be understood as follows:

  1. Egalitarianism: Equal distribution of scarce goods, regardless of individual performance.
  2. Performance-based justice: Distribution is based on performance.
  3. Needs-based justice: Meeting the basic needs of every person (e.g., food, health, education).
  4. Communist justice: Distribution according to the principle "from each according to their ability, to each according to their needs."
  5. First come, first served: Goods are allocated in the order in which needs are reported.
  6. Random principle: Equal chances of winning in the distribution of scarce goods.
  7. Authoritarian principle: Allocation by an authority figure.
  8. Procedural justice: Focus on equal procedural rules and opportunities for participation.

9. Equal opportunity: Ensuring equal starting conditions, especially through education.

   Some of these concepts conflict with each other. For example, egalitarianism and performance-based justice are hardly compatible, while other concepts, such as procedural and equal opportunity justice, can be combined.

Furthermore, understandings of justice are context-dependent and can be implemented differently depending on the region, time, and culture. It is particularly difficult to transfer these concepts to global or intergenerational scales, as implied by the principle of sustainable development (Kühne et al., 2024).

- Resilience and vulnerability: Sustainable development aims to strengthen social resilience and reduce vulnerability. Two main strategies can be distinguished here:

  1. Adaptation: Measures to cope with unavoidable changes, such as climate adaptation in construction.
  2. Prevention: Preventive measures to mitigate or prevent environmental changes, e.g., emission reduction.

These strategies can be implemented both individually (e.g., adaptation of private households) and collectively (e.g., securing critical infrastructure) (Kühne et al., 2024).

- Temporal and spatial differentiation of sustainability: The assessment of the sustainability of actions is often time- and location-dependent. What is considered sustainable in one place may not be viable elsewhere. This can be seen, for example, in land use: in certain regions, extensive livestock farming may be more sustainable than arable farming, for example, in dry savannas or mountain regions that are prone to erosion.

  Temporary aspects also play a role. For example, temporally changing forms of cultivation such as field-pond farming can be more ecologically sustainable than permanent monocultures, depending on the respective environmental conditions and cultivation cycles (Ekart, 2005; Grunwald & Kopfmüller, 2012; Pufé, 2017; Rogers et al., 2012; von Hauff & Kleine, 2009).

The preceding remarks have reconstructed the historical development of the concept of sustainability as well as the genesis and political anchoring of the SDGs as a global system of goals. It has become clear that the discourse on sustainable development cannot be reduced to a coherent interpretation but is rather shaped by competing understandings, cultural semantics, and political interests. This diversity is reflected not only in the interpretation of the SDGs themselves but also in the way sustainability is understood as a normative framework, operationalized, and linked to concrete expectations for action. The resulting tensions provide the starting point for a systematic examination of the diverse forms of criticism that can be articulated against the SDG architecture—including not only internal criticism, contextualizations, and ideological objections, but also metatheoretical classifications, pragmatic questions, and an analysis of intended and unintended consequences.

The following analysis focuses on the application and contextualization of the forms of criticism differentiated in Sect. 2.2.2. In particular, the neo-pragmatic approach—with its focus on contingent vocabularies, situated claims to validity, and the expansion of "we" in Rorty's sense—opens up a productive interpretive framework: it allows criticism to be understood not as a purely destructive act, but as an impulse for revising and improving discursive practices. In this sense, the critical examination of the SDGs should be understood not only as an analysis of normative inconsistencies or structural gaps but also as a contribution to the reflexive further development of the vocabulary of sustainable development itself.

References

Brühl, T., & Rosert, E. (2014). *Die UNO und Global Governance*. Springer Fachmedien Wiesbaden. https://doi.org/10.1007/978-3-658-00143-8

Carson, R. (1962). *Silent Spring*. Mifflin.

Diamond, J. M. (2011). *Collapse: How Societies Choose to Fail or Survive*. Penguin.

Ekardt, F. (2005). *Das Prinzip Nachhaltigkeit: Generationengerechtigkeit und globale Gerechtigkeit*. Beck.

Förstner, U., & Köster, S. (2018). *Umweltschutztechnik*. Springer Berlin Heidelberg. https://doi.org/10.1007/978-3-662-55163-9

Gethmann, C. F. (2009). *Untersteht alle Forschung dem Prinzip des Fallibilismus, nur die Klimaforschung nicht?* Akademie-Brief. https://www.ea-aw.de/fileadmin/downloads/Newsletter/NL_0087_022009.pdf

Grober, U. (2013). Von Freiberg nach Rio – Carlowitz und die Bildung des Begriffs 'Nachhaltigkeit'. In G. A. Jahn (Ed.), *Die Erfindung der Nachhaltigkeit: Leben, Werk und Wirkung des Hans Carl von Carlowitz* (pp. 13–30). oekom-Verl.

Grunwald, A., & Kopfmüller, J. (2012). *Nachhaltigkeit: Eine Einführung* (2., aktualisierte Auflage). *Campus »Studium«*. Campus.

KOM. (2011). *Mitteilung der Kommission an das Europäische Parlament, den Rat, den europäischen Wirtschafts- und Sozialausschuss und den Ausschuss der Regionen: Rio+20: Hin zu einer umweltverträglichen Wirtschaft und besserer Governance: (363 endgültig)*.

Köhler, G. (2015). Die Millenniums-Entwicklungsziele – ein kritischer Rückblick und optimistischer Ausblick. *Vereinte Nationen: German Review on the United Nations, 63*, 243–248.

Kühne, O., Silina, I., & Berr, K. (2024). *Nachhaltige Entwicklungen und Lebenschancen*. Springer Fachmedien Wiesbaden. https://doi.org/10.1007/978-3-658-45998-7

Küster, H. (2005). *Das ist Ökologie: Die biologischen Grundlagen unserer Existenz*. Beck.

Küster, H. (2013 [1995]). *Geschichte der Landschaft in Mitteleuropa: Von der Eiszeit bis zur Gegenwart*. Beck.

Martens, J. (2018). *Zivilgesellschaft und Agenda 2030: Neue Allianzen für die sozial-ökologische Transformation?* Global Policy Forum.

Meadows, D. H., Meadows, D., Randers, J., & Behrens, W. W., III. (1972). *The Limits to Growth: A Report for the Club of Rome's Project on the Predicament of Mankind*. Universe Books.

Michelsen, G. (2012). Grundlagen nachhaltiger Entwicklung.

Michelsen, G. (Ed.). (2017). *forum hlz. Die deutsche Nachhaltigkeitsstrategie: Wegweiser für eine Politik der Nachhaltigkeit* (2nd ed.). Hessische Landeszentrale für politische Bildung.

Mishan, E. J. (1967). *The Costs of Economic Growth*. Staples Press.

Pradhan, P., Costa, L., Rybski, D., Lucht, W., & Kropp, J. P [Jürgen P.] (2017). A Systematic Study of Sustainable Development Goal (SDG) Interactions. *Earth's Future*, *5*(11), 1169–1179. https://doi.org/10.1002/2017EF000632

Pradhan, P., Reusser, D. E., & Kropp, J. P. [Juergen P.] (2013). Embodied Greenhouse Gas Emissions in Diets. *PLOS One.* Advance online publication. https://doi.org/10.1371/journal.pone.0062228

Pufé, I. (2017). *Nachhaltigkeit* (3., überarbeitete und erweiterte Auflage). *UTB: Vol. 8705.* UVK.

Radkau, J. (1983). Holzverknappung und Krisenbewußtsein im 18. Jahrhundert. *Geschichte Und Gesellschaft, 9*(4), 513–543. http://www.jstor.org/stable/40185322

Radkau, J. (2000). *Natur und Macht: Eine Weltgeschichte der Umwelt.* Beck.

Radkau, J. (2017). *Geschichte der Zukunft: Prognosen, Visionen, Irrungen in Deutschland von 1945 bis heute.* Carl Hanser Verlag.

Rogall, H. (2008). *Ökologische Ökonomie: Eine Einführung* (2., überarb. und erw. Aufl.). VS Verl. für Sozialwiss.

Rogers, P. P., Jalal, K. F., & Boyd, J. A. (2012). *An Introduction to Sustainable Development.* Earthscan.

Schenk, W. (2011). *Historische Geographie. Geowissen kompakt.* WBG.

Schwartz, S. (2016). *Nachhaltigkeit als Komplexitätsfalle* [Dissertation, Universitäts- und Landesbibliothek Münster]. K10plus. http://nbn-resolving.de/urn:nbn:de:hbz:6-67209493759

Sieferle, R. P. (1984). *Fortschrittsfeinde? Opposition gegen Technik und Industrie von der Romantik bis zur Gegenwart. Die Sozialverträglichkeit von Energiesystemen: Vol. 5.* Beck.

Spindler, E. A. (2012). *Geschichte der Nachhaltigkeit: Vom Werden und Wirken eines beliebten Begriffes.* https://www.nachhaltigkeit.info/media/1326279587phpeJPyvC.pdf

UNCED. (1992). *Agenda21.* UNCED.

United Nations. (2015). *Resolution Adopted by the General Assembly on 25 September 2015: 70/1. Transforming Our World: The 2030 Agenda for Sustainable Development.* https://www.un.org/en/development/desa/population/migration/generalassembly/docs/globalcompact/A_RES_70_1_E.pdf

Unmüßig, B. (2003). Raubbau an natürlichen Ressourcen und Elemente einer ökologischen Gestaltung der Globalisierung. In M. Massarrat, U. Rolf, & H.-J. Wenzel (Eds.), *Bilanz nach den Weltgipfeln: Perspektiven für Umwelt und Entwicklung; [Rio de Janeiro 1992 - Johannesburg 2002]* (1. Aufl., pp. 11–21). OEKOM VERLAG; ökom-Verl.

Vereinte Nationen. (2015). *MDG Report 2015.* Vereinte Nationen.

von Carlowitz, H. C. (2022 [1732]). *Sylvicultura oeconomica: oder Haußwirthliche Nachricht und Naturmäßige Anweisung zur Wilden Baum-Zucht* (J. Hamberger, Ed.). oekom.

von Hauff, M. (2021). *Nachhaltige Entwicklung: Grundlagen und Umsetzung* (3., überarbeitete und erweiterte Auflage). De Gruyter Oldenbourg. https://www.degruyter.com/books/9783110722369

von Hauff, M., & Kleine, A. (2009). *Nachhaltige Entwicklung: Grundlagen und Umsetzung.* Oldenbourg Wissenschaftsverlag. https://doi.org/10.1524/9783486850949

Weiland, S. (2007). *Politik der Ideen: Nachhaltige Entwicklung in Deutschland, Großbritannien und den USA.* VS Verlag für Sozialwissenschaften Wiesbaden.

Wölkner, S. (2019). Agenda 2030: *Mut zur Nachhaltigkeit!* Konrad Adenauer Stiftung.

World Commission on Environment and Development. (1987). *Our Common Future.* World Commission on Environment and Development.

Chapter 7
Criticism of the SDGs

Based on a theory-driven critical framework, the following section examines the SDGs. It goes beyond common criticism, which primarily focuses on the structure of the goals, internal contradictions, and weaknesses in implementation (see Fukuda-Parr, 2016; Hickel, 2017; Pogge & Sengupta, 2016). Less attention has been paid to the interrelationships between the goals and their theoretical prerequisites.

The aim is to address this analytical gap by also considering those aspects that are not explicitly addressed in the SDGs (cf. Hard, 2008; Kühne, 2024; Kühne et al., 2024a).

7.1 Internal Criticism

The United Nations sees Agenda 2030 as a global roadmap for promoting peace, prosperity, and environmental sustainability (United Nations & UNRIC—United Nations Regional Information Centre, n.d., n.p.). Although the preamble refers to the classic three-pillar model (economy, social affairs, environment), the resolution (United Nations, 2024) lacks a clear definition of sustainable development. The SDGs themselves appear partially contradictory and imprecise: some goals overlap, while key issues—such as financing or the compatibility of ecological sufficiency with economic growth—remain unresolved. SDG 13 in particular is implicitly reflected in many other goals without this being strategically reflected (Kühne et al., 2024b).

A fundamental conflict of objectives arises from the idea of achieving global justice (SDG 10) through economic growth in the Global South, even though this hardly seems realistic under current environmental conditions. The necessary redistribution of wealth is not addressed, nor are the political conflicts that such processes can give rise to—especially regarding the power resources of the Global North (in

This chapter essentially summarizes our remarks in Kühne et al. (2024).

O. Kühne et al., *Sustainable Development and Life Chances*, Neopragmatic Horizontal Geographies 3, https://doi.org/10.1007/978-3-032-20182-9_7

the sense of Bourdieu, 1979 [French original 1972]). Geopolitical tensions and the productive function of conflicts (cf. Dahrendorf, 1972, 1994) are also ignored, which makes the agenda appear largely conflict-free and utopian. Added to this is the contradiction between the universalist claim (especially SDG 17) and the national differentiation provided for in point 59 (United Nations, 2024). Implementation is top-down, with responsibility shifted to subnational levels and individuals, without clear mechanisms for adaptation in the event of unexpected events such as pandemics or geopolitical crises (Kühne et al., 2024b).

The indicator system also proves problematic: first, concrete indicators are contrasted with vague goals. Second, there is often no logical derivation of the indicators from the goals—as is the case with SDG 4 on education. Third, there is a lack of reliable data collection at the local level, partly due to data protection. Fourth, the collection of indicators itself generates opportunity costs, as resources are spent on administration rather than on achieving goals (Kühne et al., 2024b).

7.2 Criticism from Contextualization

As already mentioned, the concept of sustainability in Agenda 2030 remains largely vague, apart from the three-pillar model. This makes it difficult to conduct a well-founded contextual critique, for example, regarding normatively presupposed notions of justice. Concepts of needs-based justice, equal opportunity, egalitarianism, procedural justice, and performance-based justice were therefore used for the analysis.

Many SDGs (including SDG 1 [No Poverty], SDG 2 [Zero Hunger], and SDG 3 [Good Health and Well-being]) can be clearly assigned to needs-based justice: they relate to securing basic needs such as food, health, water, energy, infrastructure, and the environment. SDG 4 (Quality Education) and SDG 5 (Gender Equality) also include equal opportunity and egalitarianism, for example, through the right to equal access to education. Procedural justice is evident in SDG 12 (Responsible Consumption and Production), 16 (Peace, Justice, and Strong Institutions), and 17 (Partnerships for the Goals), for example, through participatory processes or the establishment of equitable institutions. SDG 10 (Reduced Inequalities) emphasizes egalitarianism, although equal opportunities are conspicuously absent. References to performance-based justice are only found indirectly, for example, in the context of efficient resource use (Kühne et al., 2024b).

Concepts of justice such as communism, the first-come-first-served principle, the random principle, or the authoritarian principle are not considered—which seems logical given the emphasis on participation (especially in SDG 16). Nevertheless, a strongly needs-based understanding dominates, which corresponds to the lower levels of Maslow's hierarchy of needs (Maslow, 1977)—i.e., it focuses primarily on physiological, safety, and social needs (Kühne et al., 2024b). The top-down nature of many SDGs is striking, especially in SDG 17. This perspective emphasizes government control (government) more than social self-organization (governance) (see also

Addink, 2019; Benz, 2004; Benz et al., 2007; Blumenthal & Bröchler, 2006; Mayntz, 2006). The positioning of SDG 17 immediately after SDG 16 reinforces the impression that participatory processes are being replaced by centralized instruments when in doubt.

Furthermore, cultural and regional differences are hardly addressed. The universalistic approach often reflects the perspectives of the Global North, marginalizing the perspectives and needs of the Global South (Hickel, 2017; Ziai, 2016). Differences in government capacity, access to resources, or dependence on natural resources—for example, in the maritime sector—are largely ignored. The increasing financialization of the SDGs should also be viewed critically. Sustainability is increasingly being marketed as an investment product—e.g., through green bonds or impact investing—which can shift the focus from social goals to returns on capital (Bracking & Leffel, 2021). This increases the risk that economic interests will dominate the agenda. Another blind spot lies in the agenda's cognitivist orientation. Emotional, affective, or aesthetic connections to fellow human beings, the landscape, or nature, which often guide our actions (e.g., Forkel & Grimm, 2014; Kühne, 2013; Schemel, 2004; Wagner, 1997), are hardly considered.

In summary, it can be said that the SDGs convey a strongly predetermined, deterministic understanding of sustainable development. Local, cultural, and social contexts appear less as a resource than as an obstacle to be overcome. This means that countries in the Global South in particular are bound by a development agenda that appears to offer no alternatives (Kühne et al., 2024b).

7.3 Ideological Criticism

Concepts of justice reflect fundamental ideological attitudes (Kühne et al., 2021, 2022). The following critique of the SDGs draws on the concepts of egalitarianism, communist principles, performance-based justice, the first-come, first-served principle, equal opportunity, and procedural justice, omitting the principle of needs-based justice, which is already central to all goals (Kühne et al., 2024b). The random principle and the authoritarian principle are excluded because they contradict the goal of orderly, self-determined development.

The communist principle can be considered a variant of egalitarianism, supplemented by the consideration of individual performance. Both call for more far-reaching social redistribution—not only legally, but also economically and culturally. The SDGs lack corresponding approaches; neither global power structures nor historical inequalities, such as those caused by colonialism, are addressed. There is no critical examination of capitalist growth logic, nor are there any concrete demands for sufficiency or comprehensive redistribution. From this perspective, the SDGs therefore fail to meet the requirements for structurally just conditions (Kühne et al., 2024b).

From the perspective of performance-based justice—which emphasizes the relationship between contribution and entitlement—the dominance of needs-based justice appears problematic. It relies on redistribution, regardless of individual performance, which can reduce performance incentives and limit innovation. The first-come, first-served principle exacerbates this view: those who act faster should benefit more. However, the SDGs largely ignore such dynamics. A stronger emphasis on the speed of achieving goals, on the other hand, could create additional incentives.

Equal opportunity and procedural justice are closely linked, as fair starting conditions are tied to transparent and accessible procedures. From this perspective, it is criticized that the SDGs mitigate existing disadvantages but do not necessarily promote individual empowerment. Furthermore, needs-based outcomes do not automatically guarantee fair procedures—arbitrary distributions would be possible. Both concepts of justice aim to overcome the disadvantages of egalitarianism (leveling) and performance-based justice (unequal starting points) (Kühne et al., 2024b).

In the SDGs, equal opportunity is almost exclusively reflected in SDG 4 (Quality Education), although ideally it would be a cross-cutting goal. Similarly, procedural justice is limited to SDG 16 (Peace, Justice, and Strong Institutions), although fair procedures are fundamental to all target areas. Overall, it is clear that the SDGs primarily follow a narrow understanding of justice focused on basic needs, which lacks a broader, structure-oriented perspective.

7.4 Metatheoretical Criticism

The SDGs are largely based on a positivist understanding of the world that favors scientific logic. It is dominated by an understanding of 'truth' that can be derived from measuring, weighing, counting, and modeling based on these measurements. This logic also shapes the structure of the SDGs: measures are derived linearly from problems, and their effectiveness is to be verified using indicators (Kühne et al., 2024b).

Such an approach largely ignores the complexity and contingency of social conditions. It reduces social processes to causal models in which even the relationships between humans and the environment are recorded exclusively in quantitative terms. This creates the impression that even intersubjective or human-nonhuman relationships can be controlled and regulated. However, the strong top-down orientation of the SDGs—in the sense of classic government logic—presupposes functioning global structures. If these are not in place, the claim to control appears more like overconfidence than a realistic strategy (Kühne et al., 2024b).

This complexity-reducing worldview is characterized by a reliance on dispositional knowledge: what needs to be done seems to be known, as do the how and why. However, it fails to reflect that, under changed socio-ecological conditions, orientational knowledge would be more appropriate. This carries the risk of unintended consequences, which will be discussed in the next section.

7.5 Criticism of the Consequences and Unintended Side Effects of Research

The SDGs exhibit an implicit teleology—not at the conceptual level, but through their focus on quantified targets and indicators. This makes it difficult to adapt to unforeseeable developments and rules out alternative solutions from the outset. Feedback loops that examine not only target achievement but also target appropriateness are largely absent. This linearity carries risks of unintended side effects, as future needs and conditions—for example, under the influence of climate change or growing knowledge bases—are difficult to anticipate. The positivist thinking inherent in the SDGs, with its universalistic and technologically deterministic approach, can overlook regional, cultural, and ecological contexts, thereby exacerbating social and ecological tensions (Kühne et al., 2024b).

In extreme cases, new inequalities arise, for example, when local societies are economically and culturally oriented toward unsustainable resources such as fossil fuels. This highlights the need for an ethical relationship between goals, measures, and indicators—including cognitive and emotional perspectives. Such reflection has been lacking in the concept of the SDGs (Kühne et al., 2024b).

7.6 Pragmatic Criticism

The vision of sustainable development remains vague in the 2030 Agenda, and its communication often appears simplistic (United Nations and UNRIC n.d.). Potential conflicts—such as those arising from necessary lifestyle restrictions in the Global North—are hardly addressed, let alone solutions offered. Recommendations for action remain overly simplistic, such as blanket recommendations for vegan diets without reflecting on their side effects on biodiverse land use systems such as alpine farming (Kühne et al., 2024b).

The top-down logic of the SDGs transfers responsibility to nation-states, which intervene directly or indirectly in everyday life through legislation. Implementation often follows a scientific-causal line of reasoning, in which research findings are used to derive policy measures that are then translated into individual behavior through legislation—without considering the complexity of social relationships or the desire for self-determination (Kühne et al., 2024b). Individual implementation remains limited. Suggestions such as action weeks or role-playing games with characters such as "veggie advocates" seem banal or infantilizing, which can be met with rejection, especially among adults. In addition, the question arises as to how realistic long-term motivation is when individual sacrifices (renunciation) are made early on, while benefits only materialize later—comparable to discounting in the financial world: high initial costs are discouraging when returns (benefits) are distant and uncertain (Kühne et al., 2024b).

This problem is not new: the philosophical discussion about akrasia (cf. Seebaß, 2005) shows that insight does not necessarily lead to action. Norms need to be anchored in society—moral conviction, not mere command. This also echoes Böckenförde's dictum: a liberal state thrives on conditions that it cannot guarantee itself (Böckenförde, 1976, p. 60). Nevertheless, some voices—such as Carolin Emcke (Schneider, 2024)—call for authoritarian measures to enforce "correct" behavior. This contradicts liberal democracy and fails to recognize that sustainable behavior must be voluntary and conviction-based. Resorting to paternalism or moral pressure when there is no effect is a well-known danger in democracy and totalitarianism research (exemplary references in Kühne et al., 2021, 2022).

In addition, there are conflicting goals: sustainable measures often conflict with individual or collective interests. In pluralistic societies, there is no overarching principle for resolving these conflicts—instead, the well-known free rider problem arises. Those who act reasonably pay the costs, while others benefit without contributing. According to rational choice theory (Diefenbach, 2009), this reinforces the tendency to defect, i.e., to reject collective responsibility, through the prisoner's dilemma (Axelrod, 2000), with negative consequences for everyone (Kühne et al., 2024b).

7.7 Summary

The above discussion reveals a fundamental contradiction in the SDGs: the goals are ambitious, but some of them remain unspoken and are presented in an inconsistent and overly simplistic framework. This makes it difficult to derive concrete fields of action and measures and, above all, to evaluate the achievement of the goals. Although the SDGs follow a positivist or even scientistic understanding of the world, this claim itself is not fulfilled. Instead, imprecise indicators, some of which are impossible to collect, lead to high administrative costs and opportunity costs that hinder actual implementation. Furthermore, the SDGs lack coherence in terms of content. Goals contradict each other, are formulated twice, or appear eclectic because there is no common understanding of sustainable development. Instead, a nave, utopian belief in progress dominates, ignoring social conflicts—even though, as Dahrendorf (1972) emphasizes, these are inherent in any social transformation. From a pragmatic perspective, it is particularly problematic that profound changes in everyday life are demanded without recognizing that this devalues existing symbolic capital and creates new pressure to adapt.

There is also a lack of crisis resilience: neither pandemics, wars, natural disasters, nor geopolitical conflicts are systematically considered. Instead of flexible approaches, a linear logic of progress remains in place, which relies on compliance with quantitative indicators. The reference to World 2—to people's inner lives—remains rudimentary, and the relationship between the individual and society is strongly instrumental and top-down. As shown, an understanding of needs-based justice dominates, focusing on the lower levels of Maslow's hierarchy of needs. This may be appropriate for many regions with lower symbolic capital, but it remains unclear what sustainable development should look like once these basic needs are met.

The inconsistencies identified here and the limited operationalizability of the SDGs make it clear that an integrative understanding of sustainability is needed. A corresponding conceptual proposal is developed in the following chapter.

References

Addink, H. (2019). *Good Governance: Concept and Context*. Oxford University Press.

Axelrod, R. M. (2000). *Die Evolution der Kooperation*. Oldenbourg.

Benz, A. (Ed.). (2004). *Governance – Regieren in komplexen Regelsystemen: Eine Einführung*. VS Verlag für Sozialwissenschaften.

Benz, A., Lütz, S., Schimank, U., & Simonis, G. (Eds.). (2007). *Handbuch Governance: Theoretische Grundlagen und empirische Anwendungsfelder*. VS Verlag für Sozialwissenschaften.

Blumenthal, J. V., & Bröchler, S. (Eds.). (2006). *Von Government zu Governance. Analysen zum Regieren im modernen Staat*. LIT.

Bourdieu, P. (1979 [frz. Original 1972]). *Entwurf einer Theorie der Praxis auf der ethnologischen Grundlage der kabylischen Gesellschaft*. Suhrkamp.

Böckenförde, E.-W. (1976). *Staat, Gesellschaft, Freiheit: Studien zur Staatstheorie und zum Verfassungsrecht. Suhrkamp-Taschenbuch Wissenschaft: Vol. 163*. Suhrkamp.

Bracking, S., & Leffel, B. (2021). Climate Finance Governance: Fit for Purpose? *Wiley Interdisciplinary Reviews: Climate Change, 12*(4), Article e709. https://kclpure.kcl.ac.uk/portal/en/publications/climate-finance-governance-fit-for-purpose

Dahrendorf, R. (1972). *Konflikt und Freiheit: Auf dem Weg zur Dienstklassengesellschaft*. Piper.

Dahrendorf, R. (1994). *Der moderne soziale Konflikt: Essay zur Politik der Freiheit*. dtv.

Diefenbach, H. (2009). Die Theorie der Rationalen Wahl oder „Rational Choice"-Theorie (RCT). In D. Brock, H. Diefenbach, M. Junge, R. Keller, & D. Villányi (Eds.), *SpringerLink Bücher. Soziologische Paradigmen nach Talcott Parsons: Eine Einführung* (pp. 239–290). VS Verlag für Sozialwissenschaften. https://doi.org/10.1007/978-3-531-91454-1_5

Forkel, J. A., & Grimm, M. (2014). Die Emotionalisierung durch Landschaft oder das Glück in der Natur. *Sozialwissenschaften und Berufspraxis, 37*(2), 251–266.

Fukuda-Parr, S. (2016). From the Millennium Development Goals to the Sustainable Development Goals: Shifts in Purpose, Concept, and Politics of Global Goal Setting for Development. *Gender & Development, 1*, 43–52. https://doi.org/10.1080/13552074.2016.1145895

Hard, G. (2008). Der Spatial Turn, von der Geographie her beobachtet. In J. Döring & T. Thielmann (Eds.), *Spatial Turn: Das Raumparadigma in den Kultur- und Sozialwissenschaften* (pp. 263–316). transcript. https://doi.org/10.14361/9783839406830-011

Hickel, J. (2017). *The Divide: A Brief Guide to Global Inequality and Its Solutions*. William Heinemann.

Kühne, O. (2013). Landschaft – ein emotionales Konstrukt: Das ästhetische Erleben von Landschaften zu Zeiten der regenerativen Energiegewinnung. In DGGL (Ed.), *Energielandschaften: Geschichte und Zukunft der Landnutzung. DGGL-Jahrbuch 2013* (pp. 17–20). Callwey.

Kühne, O. (2024). *Redescribing Horizontal Geographies: A Neopragmatist Approach to Spatial Contingency, Complexity, and Relationships*. Springer International.

Kühne, O., Berr, K., & Jenal, C. (2022). *Die geschlossene Gesellschaft und ihre Ligaturen: Eine Kritik am Beispiel 'Landschaft'*. Springer VS.

Kühne, O., Berr, K., Schuster, K., & Jenal, C. (2021). *Freiheit und Landschaft: Auf der Suche nach Lebenschancen mit Ralf Dahrendorf*. Springer.

Kühne, O., Koegst, L., & Berr, K. (2024a). *Oilscapes of Louisiana: Neopragmatic Reflections on the Ambivalent Aesthetics of Landscape Constructions.* Springer VS. https://doi.org/10.1007/978-3-658-43396-3

Kühne, O., Silina, I., & Berr, K. (2024b). *Nachhaltige Entwicklungen und Lebenschancen.* Springer Fachmedien Wiesbaden. https://doi.org/10.1007/978-3-658-45998-7

Maslow, A. H. (1977). *Motivation und Persönlichkeit.* Walter.

Mayntz, R. (2006). From Government to Governance: Political Steering in Modern Societies. In F. Rubik & D. Scheer (Eds.), *Governance of Integrated Product Policy: In Search of Sustainable Production and Consumption* (pp. 18–25). Routledge.

Pogge, T., & Sengupta, M. (2016). Assessing the Sustainable Development Goals from a Human Rights Perspective. *Journal of International and Comparative Social Policy, 32*(2), 83–97.

Schemel, H.-J. (2004). Emotionaler Naturschutz – zur Bedeutung von Gefühlen in naturschutzfachlichen Entscheidungsprozessen. *Natur und Landschaft, 79*(8), 371–378.

Schneider, A. (2024). *Es gibt nur eine Wahrheit.* https://www.welt.de/debatte/kommentare/plus251917114/Demokratie-Verachtung-Es-gibt-nur-eine-Wahrheit.html

Seebaß, G. (2005). Akrasie. In J. Mittelstraß (Ed.), *Enzyklopädie Philosophie und Wissenschaftstheorie* (2., neubearbeitete und wesentlich ergänzte Auflage, pp. 59–63). Metzler.

United Nations. (2024). *Transforming Our World: The 2030 Agenda for Sustainable Development.* https://sdgs.un.org/2030agenda

Vereinte Nationen, & UNRIC—Regionales Informationszentrum der Vereinten Nationen. (o.J.). *Ziele für Nachhaltige Entwicklung.* https://unric.org/de/17ziele/

Wagner, J. M. (1997). Zur emotionalen Wirksamkeit von Kulturlandschaft. In W. Schenk, K. Fehn, & D. Denecke (Eds.), *Kulturlandschaftspflege: Beiträge der Geographie zur räumlichen Planung* (pp. 59–66). Borntraeger.

Ziai, A. (2016). *Development Discourse and Global History: From Colonialism to the Sustainable Development Goals.* Routledge.

Chapter 8
Life Chances and Sustainable Development

In the previous chapters, we dealt with a concrete critique of the SDGs after first presenting and explaining the critical framework, neopragmatism as a theoretical-conceptual framework, the possible balancing of the individual and society based on the concept of life chances and considering the dual structure of individual striving for options and social ties, and the concept of sustainability in its historical contexts. In order to gain an integrative understanding of sustainability and, if possible, to conceptualize and operationalize it, the aforementioned aspects must be related to each other in their diversity and synthesized. As shown in the previous chapters, reductionist models should also be avoided in this question. The integrative perspective is, on the one hand, Popper's model of the three worlds and, on the other hand, a metatheoretical perspective that focuses more on the development of orientational knowledge than on dispositional knowledge, as is the consequence for the implementation of the SDGs. In this chapter, we will first address fundamental considerations regarding life chances and sustainable development before turning to conceptual considerations.

8.1 Fundamental Considerations

The question is how the concept of sustainable development can be reformulated or redescribed in such a way that it is compatible with the possibility of preserving, enabling, and increasing life chances and does not instead, as has already become clear in some of the previous analyses, potentially destroy, render impossible, or prevent life chances.

Karl Popper's theory of three worlds (Popper, 1959, 1979; Popper & Eccles, 1977) offers a potential for differentiation and thus analysis that goes beyond what has been said so far. The concept of sustainable development can be assigned firstly to World 1, more specifically to animate and inanimate matter (abiotic and biotic); secondly

O. Kühne et al., *Sustainable Development and Life Chances*, Neopragmatic Horizontal Geographies 3, https://doi.org/10.1007/978-3-032-20182-9_8

to World 3 (specifically in cultural, social, political, and economic aspects); and thirdly to World 2 (specifically to individuals striving for sustainability). Both the levels of abiotic and biotic matter and the connection between World 3 and World 2 are understood as emergent phenomena: without abiotic matter, no biotic matter is possible in World 1; and without this emergent World 1, no co-evolving Worlds 2 and 3 are possible. The emergence levels of worlds 2 and 3 are explained in the context of discussions about the 'body-soul' or 'body-mind' problem by theories of non-reductive or reductive physicalism (Brüntrup, 1996; Zoglauer, 1998). World 3 cannot be limited to cultural or social aspects alone but must be differentiated into the social sub-areas of economy, politics, social community, and culture, as these have their own distinct logics, and therefore a simple subsumption under the concept of the 'social'—as in the three-pillar model of sustainable development—appears analytically undercomplex (Bourdieu, 1989, 2002; Kneer, 2004; Luhmann, 1984, 1986; Parsons, 1951). The necessity and significance of this expansion are particularly evident in the field of politics, as, both from the perspective of the open society (Popper, 2011 [1945]) and conflict theory, it plays a central role in enabling and securing individual life chances. Not only do World 2 and World 3 interact, but these different fields also interact with each other through the mediation of World 2 and, in some cases, hybridize.

The above explanations have also made it clear that World 2 plays a central role not only in Popper's three-world theory but also in his reflections on the 'open society' and in Dahrendorf's concept of life chances—and not only in these concepts themselves but also in an integrative understanding of sustainable development, as proposed by us. More specifically, the individual is a constitutive part of the biotic and abiotic World 1, participates in the cultural, social, political, and economic processes of World 3, and can intervene in Worlds 1 and 3 through these processes as well as through his or her physicality (World 1). This central position of the individual in the interweaving of the three worlds and their differentiations is visualized in Fig. 8.1.

In order for life chances to be maximized at all, certain prerequisites are necessary. First of all, adequate participation in the structures and processes of worlds 1 and 3 is essential. The term'adequate participation'" means acquiring and mastering important forms of knowledge (see examples of forms of knowledge: Abel, 2015). First, there is the transfer of explanatory knowledge, orientational knowledge, and action knowledge. Explanatory knowledge embodies an understanding of natural and social processes and knowledge of their interactions. Orientational knowledge embodies the ability to reflect on the meaningfulness and justifiability of goals and purposes that are intended to provide guidance for action, for which suitable means can then be provided if they are accepted, whereby the choice of means should also be reflected upon and justified. With regard to moral goals and purposes, the ability to reflect ethically is addressed. This means that *before* applying "general ethical imperatives or a correspondingly justified right" (Hubig, 2015, p. 92) to areas of action (here: ethical to moral ligatures), it is necessary that actors "were able to *orient themselves in advance* regarding the quality of the goals and implementation options and that this *orientation* took place" (Hubig, 2015, p. 92). This refers to "reflexive orientation" as opposed to "transitive orientation" (Hubig, 2015, p. 92), which aims to

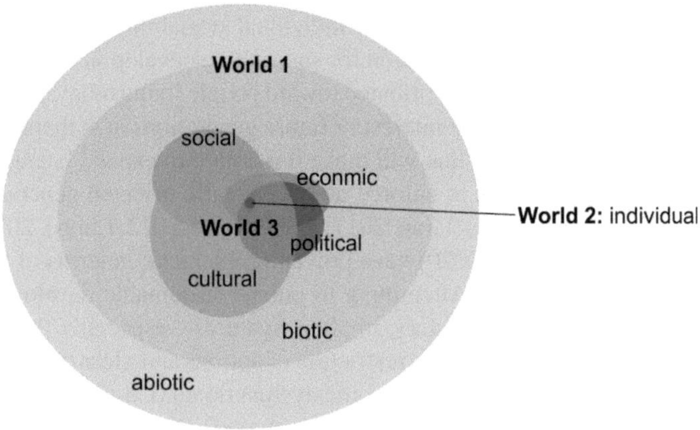

Fig. 8.1 Representation of the dimensions of sustainable development in relation to Popper's three worlds. The transparency of the individual areas (biotic and further inside) symbolizes the interweaving and hybridization of the different dimensions up to the individual, who is embedded in all of them (*Source* Kühne et al., 2024, p. 77)

provide fixed "orientation" (Hubig, 2015, p. 96). Before actors can accept appropriate guidance in the context of sustainable development, they should be able to orient *themselves*. However, this requires the ability to engage in ethical reflection in the first place and is ultimately linked to moral competencies in everyday life (Lohmann et al., 2024, pp. 91–93) and/or reflective abilities that have been developed and trained through education. Action knowledge is gained and applied on the basis of explanatory and orientation knowledge and with the inclusion of uncertainty, enabling action decisions to be made. Secondly, 'appropriate participation' means the opportunity to act as an individual (World 2) in Worlds 1 and 3: Specifically, this means being able to engage in the struggle for appropriate responses to challenges in relation to World 3, and being able to shape parts of World 1 independently according to one's own ideas (developed in interaction with Worlds 2 and 3).

8.2 Conceptual Considerations

Despite massive information and 'awareness' campaigns over the past decades, sustainable development is by no means a 'sure-fire success'. The idea that providing information and raising awareness about ecological, social, and economic issues alone would be enough to fundamentally change people's lives in favor of sustainable development has proven to be largely naive and unrealistic, and therefore not very viable. In research, this problem, which has been known since ancient times, is referred to as the 'akrasia' problem (cf. Seebaß, 2005), namely the gap between insight and action (see Sect. 7.6). People are often unwilling to voluntarily enter into new ligatures and forgo individually available latent options or manifest options in order to create latent options whose future individual availability is not clear to them. Latent options are a necessary condition for sustainable development geared toward maximizing life chances, which is oriented toward people living today, but especially toward future generations. In the interest of future generations, it is therefore important to create conditions today that will make it possible to expand latent options in the future. However, attempts to enforce new, externally oriented general ligatures through moralization strategies (Grau, 2017; Kühne et al., 2022; Lübbe, 2019; Stegemann, 2018; Strohschneider, 2020) have proven problematic in terms of increasing the number of latent options. Attempting to pursue sustainable development with resentment-driven moral, involuntary, implicit ligatures—especially through physical mediation—ultimately leads to a restriction of options and a loss of the possibility of conflict resolution. Conflicts then become dysfunctional (Coser, 2012 [1956]) and counterproductive with regard to the 'implementation' of sustainable development goals, insofar as they cannot exploit their functional potential in terms of the goal of expanding life chances. In addition, ligatures are designed for the long term; they impose permanent demands on people without offering new options in the short term. For pragmatic reasons of feasibility, sustainable development should therefore focus on options (which can be offered and seized in the short term) rather than on permanent and long-term changes to ligatures. In view of the unpleasant practical consequences, the idea of motivating people to give up some of their individually available latent options (*freedom*) by creating abstract options (*liberty*) seems rather unrealistic. If the range of available options is restricted by ligatures that are directed externally at the individual, this almost inevitably provokes opposition. This opposition can give rise to social conflicts that can become highly intense and violent. These considerations lead to the following conclusion: the success of sustainable development anchored in large parts of society depends on increasing individually available latent options. The expansion of latent options alone is not an equivalent substitute for lost individually available options. Instead, people need to be empowered to take advantage of the options that have been created. It is not enough, for example, to expand public transport; it must also be easily accessible to as many people as possible and operate at times of the day when there is actually and increasingly individual demand.

This concretization of the understanding of options in relation to sustainable development can illustrate that, although the increase in latent options initially increases the chance of promoting solutions to problems relating to sustainable development in the future, the more concrete the perception of the options becomes, the more conflict-prone it becomes. This already applies to individually available latent options, because these can compete individually with (scarce) individually available latent options of other people. This applies in particular to manifest options, namely when a person chooses an option that may then no longer be available to another person or produces unintended side effects that create (involuntary) constraints for other people. In this way, competition for options gives rise to different conflicts with varying degrees of intensity, brutality, number of people involved, and spatial relevance (Weber et al., 2018).

This focus on life chances also has implications for the possible invocation of needs in order to justify sustainable development in a supposedly objective manner. It is often assumed that sustainable development goals are primarily aimed at fulfilling generalized (and thus often essentialistically constructed) needs. However, needs are primarily individually oriented, they are contextualized intersubjectively, and they are passively constructed. Furthermore, needs can change beyond the lower levels of Pavlov's pyramid of needs. An orientation towards life chances, on the other hand, emphasizes the active role of the individual, who is oriented towards individually available options and can give them individually relevant meaning through voluntarily entered into commitments and ethical reflections. In this sense, individual needs are actively shaped and reflected upon.

Furthermore, such an orientation towards life chances also reflects the scope of one's own claim to validity. This scope varies, especially with regard to World 1, because natural conditions such as climate, morphology, soils, or vegetation vary considerably in some cases. If such local, regional, and national (i.e., particular) differences are to be taken into account, an understanding of sustainable development geared toward increasing life chances can only proceed from a universal claim to validity at a meta-level, which must be supplemented by particularism at the object level. This differentiation also affects understandings of justice in the context of sustainable development and, once again, the question of needs. Needs-based justice (in its more passive form) is replaced at the meta-level by opportunity and procedural justice, which is geared towards equal opportunities for all. In this way, the egalitarianism propagated by many and mistakenly attributed to the object level (which is also often equated with outcome justice; cf. Krebs, 2000) is elevated to the meta-level of procedures, namely as equal opportunity to participate in procedures. At the object level, however, egalitarianism leads to the loss of latent options and thus also individually available options through the expansion of generalized, outward-looking, often moral ligatures. This separation of universalism at the meta-level and particularism at the object level also allows us to circumvent the problem of utopianism inherent in the SDGs, which understand sustainability as a teleologically defined 'end state'. In view of a constantly changing world that is always in need of useful solutions to new challenges, such a teleological conception seems to make little sense, let alone be practicable.

Furthermore, the (particular) life chances created by means of opportunity and procedural justice at the object level are always open to innovation. This also means a change in the type of ligatures that are (or must be) used: if goals are formulated in very general terms and not adapted to particular circumstances, this means that, in concrete terms, the gap must be filled with moral ligatures, as otherwise implementation would prove overly complex. Recourse to ethical ligatures, on the other hand, means that the appropriateness of the perception of latent and individually available latent options can be weighed up in accordance with immediate social and (natural) spatial contexts. Compared to a teleological top-down orientation, such a concept of sustainable development geared towards maximizing life chances does not ignore either the spatial diversity of the world or social change.

8.3 Positions

The integrative approach to sustainable development outlined above gives rise to a number of concrete positions on specific approaches or operationalizations of the understanding of sustainability that have been addressed in the course of the above discussion (see in particular Sect 6.3).

8.3.1 Procedural Understanding of Sustainability

In view of the remarks on the productivity of conflicts and the differentiated understanding of ligatures and options within the framework of the life chances concept, we reject the theoretical option of a substantive understanding of sustainability and favor a procedural understanding instead. We therefore also reject utopian ideas of a 'sustainable target society', as these tend towards totality and inevitably limit individual life chances and the development of alternative, possibly more appropriate and suitable strategies (such as scientific theories that are rejected from the outset for ideological reasons) for dealing with challenges. Accordingly, such utopianism is associated with the strengthening of moral ligatures and the moral devaluation of alternative ways of thinking and acting that do not correspond to the actions and ideas prescribed and desired by the utopian goal. In this way, options are limited. A procedural understanding of sustainability, on the other hand, seeks to expand options, ethically question moral ties, and strengthen competition for an open and appropriate approach to concrete challenges. A 'myth of finality' derived from principles of whatever kind (Beck et al., 2001), which produces moral constraints and is disseminated socially by means of 'officialization strategies' (Bourdieu, 1979 [French original 1972]),[1] is not an appropriate means of strengthening sustainable development. On the other hand, a promising approach, consistent with the above, is to focus on expanding individual options and abilities to deal with sustainability challenges responsibly, i.e., on the basis of ethical ligatures. In the (neo)pragmatic tradition, the creation and reflective use of options as possibilities for individual action are moving into the conceptual focus, replacing sustainability utopias that seek to force individual action toward this goal on the basis of moral constraints.

[1] Bourdieu (1979 [French original 1972], p. 90), defines "officialization" as a strategy for "transforming 'selfish', private, individual motives and interests [...] into altruistic, collective, publicly justifiable, in short, legitimate motives and interests."

8.3.2 *Efficiency Strategy*

Given our understanding of maximizing individual life chances, the efficiency strategy also appears more promising than the consistency and sufficiency strategies. Efficient action saves resources and thus reduces opportunity costs. Consistency strategies, on the other hand, may not increase life chances but may even reduce them by aligning with an (inefficient) development path. The sufficiency strategy is subject to even greater reservations in this regard, as it is often linked to moral constraints that are geared toward a generalization at the object level that limits options and thus life chances. The sufficiency strategy can only be understood as a positive contribution to sustainable development if the necessary individual decisions are made after weighing up options and resorting to ethical constraints. However, this is not the case if this strategy is understood as a general utopia of sufficiency.

8.3.3 *Weak Sustainability*

In view of the maximization of individual life chances, the concept of 'weak sustainability' is to be favored. We agree with this position insofar as we also agree in principle with the use of non-renewable raw materials, provided that future generations are compensated for this. The (probable) loss of life chances resulting from the current use of non-renewable resources should be overcompensated. According to this understanding, these are therefore 'things' that enable options and are not strongly determined in terms of structure and function. An example: the production of knowledge that is guided by ethical, not moral, ligatures is very likely to open up life chances for future humans. The construction of expressways, on the other hand, is geared towards current needs, and their function and structure are highly determined, so that it is highly probable that future generations will benefit less from them. The orientation towards the position of 'weak sustainability' is therefore always subject to the proviso of overcompensating for the reduction in life chances through the use of non-renewable resources with something that is likely (because the future always remains open) to open up new options for future generations in their interaction with the world (in conjunction with ethical constraints).

8.3.4 *Rejection of Collectivist Ideas*

With this world 2-centered approach, we also distance ourselves from collectivist ideas of sustainable development, as these are ultimately based on a reduction of options in favor of moral constraints. Moral constraints can be understood as an essential element of over-simplification and reduction of over-contingency: Complex causal relationships are reduced to simple normative formulas ("You must not eat animal products," "Participating in motorized private transport is evil"), contingent alternative understandings of the world and contingent options for action are excluded from discourse and morally discredited, and people who advocate this are pathologized (Berr & Kühne, 2019; Grau, 2017; Kühne, 2019).

References

Abel, G. (2015). Formen des Wissens im Wechselspiel. *Allgemeine Zeitschrift Für Philosophie, 40*(2–3), 143–160.
Beck, U., Bonß, W., & Lau, C. (2001). Theorie reflexiver Modernisierung – Fragestellungen, Hypothesen, Forschungsprogramme. In U. Beck & W. Bonß (Eds.), *Die Modernisierung der Moderne* (pp. 11–59). Suhrkamp.
Berr, K., & Kühne, O. (2019). Moral und Ethik von Landschaft. In O. Kühne, F. Weber, K. Berr, & C. Jenal (Eds.), *Handbuch Landschaft* (pp. 351–365). Springer VS.
Bourdieu, P. (1979 [frz. Original 1972]). *Entwurf einer Theorie der Praxis auf der ethnologischen Grundlage der kabylischen Gesellschaft*. Suhrkamp.
Bourdieu, P. (1989). Social Space and Symbolic Power. *Sociological Theory, 7*(1), 14–25.
Bourdieu, P. (2002). Das ökonomische Feld. In P. Bourdieu (Ed.), *Schriften zu Politik & Kultur: Vol. 3. Der Einzige und sein Eigenheim* (pp. 185–222). VSA.
Brüntrup, G. (1996). *Das Leib-Seele-Problem: Eine Einführung*. W. Kohlhammer.
Coser, L. A. (2012 [1956]). *The Functions of Social Conflict*. Taylor & Francis.
Grau, A. (2017). *Hypermoral: Die neue Lust an der Empörung* (2nd ed.). Claudius.
Hubig, C. (2015). Von der Anwendung der Ethik zur Ethik der Anwendung: Die Problematik der Bereichsethiken am Beispiel der Technikethik. In G. Gamm & A. Hetzel (Eds.), *Ethik - wozu und wie weiter?* (pp. 83–100). transcript.
Kneer, G. (2004). Differenzierung bei Luhmann und Bourdieu: Ein Theorievergleich. In A. Nassehi & G. Nollmann (Eds.), *Bourdieu und Luhmann: Ein Theorienvergleich* (pp. 25–56). Suhrkamp.
Krebs, A. (Ed.). (2000). *Gleichheit oder Gerechtigkeit: Texte der neuen Egalitarismuskritik*. Suhrkamp.
Kühne, O. (2019). Vom‚ Bösen' und‚ Guten' in der Landschaft – das Problem moralischer Kommunikation im Umgang mit Landschaft und ihren Konflikten. In K. Berr & C. Jenal (Eds.), *Landschaftskonflikte* (pp. 131–142). Springer VS.
Kühne, O., Berr, K., & Jenal, C. (2022). *Die geschlossene Gesellschaft und ihre Ligaturen: Eine Kritik am Beispiel 'Landschaft'*. Springer VS.
Kühne, O., Silina, I., & Berr, K. (2024). *Nachhaltige Entwicklungen und Lebenschancen*. Springer Fachmedien Wiesbaden. https://doi.org/10.1007/978-3-658-45998-7
Lohmann, P., Kühne, O., & Berr, K. (2024). *Landschaft und Gefühl – eine neopragmatistische Redeskription auf Grundlage der Philosophie Fichtes: Ein Beitrag zur humanistischen Geographie*. SpringerVS.
Luhmann, N. (1984). *Soziale Systeme: Grundriß einer allgemeinen Theorie*. Suhrkamp.
Luhmann, N. (1986). *Ökologische Kommunikation: Kann die moderne Gesellschaft sich auf ökologische Gefährdungen einstellen?* Westdeutscher Verlag.
Lübbe, H. (2019). *Politischer Moralismus: Der Triumph der Gesinnung über die Urteilskraft*. LIT.
Parsons, T. (1951). *The Social System*. Free Press.
Popper, K. R. (1959). *The Logic of Scientific Discovery*. Harper & Row.
Popper, K. R. (1979). Three Worlds: Tanner Lecture, Michigan, April 7, 1978. *Michigan Quarterly Review, 18*(1), 141–167.
Popper, K. R. (2011 [1945]). *The Open Society and Its Enemies*. Routledge.
Popper, K. R., & Eccles, J. C. (1977). *Das Ich und sein Gehirn*. Piper.

Seebaß, G. (2005). Akrasie. In J. Mittelstraß (Ed.), *Enzyklopädie Philosophie und Wissenschaftstheorie* (2., neubearbeitete und wesentlich ergänzte Auflage, pp. 59–63). Metzler.

Stegemann, B. (2018). *Die Moralfalle: Für eine Befreiung linker Politik*. Matthes & Seitz.

Strohschneider, P. (2020). *Zumutungen: Wissenschaft in Zeiten von Populismus, Moralisierung und Szientokratie*. kursbuch.edition.

Weber, F., Kühne, O., Jenal, C., Aschenbrand, E., & Artuković, A. (2018). *Sand im Getriebe: Aushandlungsprozesse um die Gewinnung mineralischer Rohstoffe aus konflikttheoretischer Perspektive nach Ralf Dahrendorf*. Springer VS.

Zoglauer, T. (1998). *Geist und Gehirn: Das Leib-Seele-Problem in der aktuellen Diskussion*. Vandenhoeck & Ruprecht.

Chapter 9
Redescription of the Life Chances-Oriented Understanding of Sustainability Based on Neopragmatism

In the previous chapters, we drew attention to problems associated with currently dominant understandings of sustainable development, such as strong normative claims or epistemological underdetermination. It has also become clear that established models, such as existing institutional arrangements, are struggling to adequately address the multitude of current and future challenges. Neopragmatism and a life chances orientation, however, provide theoretical approaches that allow sustainable development to be understood beyond essentialist claims to truth, teleological goals, and desires for harmony. Following on from the life chances approach and neopragmatism, social transformations depend in a particular way on how people coordinate their relationships with themselves and the world in resonant processes. Against this background, this chapter forms the core of our book, in which we draft a neopragmatic redescription of sustainable development.

9.1 A Departure from Representationalism

Sustainable development makes a normative claim to validity that must be justified. A well-known and common way of justifying this in many scientific disciplines is to understand what is to be justified (in this case, sustainable development) as an unavoidable derivation of a supposedly objective and true representation of a given world. Typical forms of this derivation are traditional theories of representation (cf. Janich, 1996), essentialism (cf. Kühne & Berr, 2021, pp. 71–74), as well as positivism and logical empiricism (cf. Kühne & Berr, 2021, pp. 86–99). Such a representational relationship can also be referred to as representationalism (Brandom, 2023; Rorty, 1981 [1979], 2023b). In contrast, we refer to the remarks in Sect. 3.1, where it was shown that notions of universal truth and irrefutable objectivity, against which hypotheses and theories can be clearly tested (Rorty, 1997 [1989]), cannot be upheld from a neopragmatic perspective. From this perspective, we therefore

O. Kühne et al., *Sustainable Development and Life Chances*, Neopragmatic Horizontal Geographies 3, https://doi.org/10.1007/978-3-032-20182-9_9

advocate viewing the concept of sustainable development not as something certain, but as an error-prone linguistic tool in the context of current challenges.

In line with the panrelationism advocated by Rorty (2023b, pp. 207–241), approaches to the world are fundamentally linguistic in nature and embedded in a dense network of relations. These relations determine what a 'thing' 'is' depending on context, interpretation, need, and linguistic formulation: Things are "what they are thanks to their relationship to other things" (Rorty, 2023a, p. 209). Language cannot therefore be separated from the world, just as, conversely, "the world cannot be separated from language" (Rorty, 2023a, p. 244). According to this basic idea, there can be no 'truth' outside of vocabularies and no unambiguously correct description of a thing: "There are only descriptions that, by relating this thing to other things, embed it in contexts that serve our current, diverse needs" (Rorty, 2023a, p. 54). Thus, humans' relationship to the outside world is "not representational, but causal. The world can be the cause of our arriving at certain beliefs as soon as we speak a language. These beliefs are developed in response to the causal stimuli of the world in order to be able to act successfully in it" (Heindl, 2023, p. 10). In this way, neopragmatism is associated with anti-objectivism and anti-essentialism as well as anti-representationalism. Language, too, is not a medium for representing external or internal reality, but is understood in the neopragmatic view as a tool (Müller, 2014; Rorty, 1991a), namely in the sense of the theory of use (cf. Leerhoff et al., 2010, pp. 50–52) that "words derive their meaning from their use in connection with other words" (Buschmeier, 2023, p. 3).

In both its classical and modern variants, the representation model assumes an instance that can decide whether representation (image, sign, interpretation) and the represented (model, signified, interpreted) actually correspond to each other. Thus, Spinoza's *monistic* model (cf. de Spinoza, 2010 [1677]; Schmid & Stoichita, 2010), which assumes that "the mind is a mirror of nature" (Brandom, 2023, p. 17), which assumes, first, the structural equality and thus correspondence of things (nature) and ideas (mind) as two inseparable manifestations of the divine substance, and second, the approximation or assimilation (adaequatio; cf. Kühne & Berr, 2021, pp. 65–66) of the finite and imperfect human mind to the infinite and perfect "mind of God" (Brandom, 2023, p. 17). René Descartes (2008 [1637]) separates nature and spirit as separate, independent entities (res extensa and res cogitans) in his dualistic theory of knowledge and introduces the concept of representation (Brandom, 2023, p. 18). According to this, the human spirit attempts to represent the world (for example, in the sciences) abstractly but 'correctly'. This representational relationship is also referred to as the 'correspondence theory of truth' (cf. exemplarily Gloy, 2004, pp. 92–167; Janich, 1996, pp. 30–39), insofar as the 'truth' of representation or depiction consists exclusively in the "agreement of knowledge with its object" (Kant, 1959 [1781], B 82). With Hegel and Heidegger, this understanding of truth can also be described as 'correctness' (Hegel, 1970 [1830]; Heidegger, 1992). But in order to assess this 'correctness'—again, 'correctly'—a 'quasi-divine' standpoint *outside* this 'correctness relationship' would be necessary, because humans are fundamentally only capable of an *internal* perspective—which is why traditional theories of truth have moved away from correspondence theories and toward coherence and consensus theories that rely on internal criteria of assessment (Kühne & Berr, 2021, pp. 66–67). In cognitive psychology and brain research, the justification of an 'external appropriateness' (Merker et al., 1998) of perception and object is referred to as the 'fallacy of the homunculus', 'i.e. the perceiving little person in the brain or in consciousness who looks at, recognizes, and judges the image of the event [...]' (and compares it with what is perceived and believes it can make a correspondingly 'correct' perceptual judgment. Such "image theories are now considered refuted" (Janich, 1996, p. 36).

However, according to Rorty, this departure from representationalist ideas does not mean denying the existence of a 'world out there' (Davidson, 2004; Prange, 2010; Rorty, 1989). What is denied is merely the claim to truth of this external world outside of descriptions: "Truth cannot be out there—cannot exist independently of the human mind—because sentences cannot exist or be out there in that way. The world is out there, but descriptions of the world are not. Only descriptions of the world can be true or false. The world itself—without the support of human descriptive activity—cannot be" (Rorty, 1989, p. 24). We are connected to the 'world out there' "not in a representational way, but in a causal way" (Heindl, 2023, p. 10). This means that an immediate understanding of 'world' (not only World 1, but also Worlds 2 and 3) remains closed to observers. However, this does not mean that Rorty does not consider World 1 to be 'real': "Thus Rorty can arguably be described as an ontological realist—he believes that there is a real, material world prior to the ways in which we happen to describe it—who is an epistemological nominalist: we can only know what this world is like under a description" (Calder, 2003, p. 37). Humans are causally connected to World 1; its "objective reality imposes itself on us through its resistance to our desires and the sometimes surprising and disappointing consequences of our actions, forcing us to adapt it to our purposes and ourselves to it" (Brandom, 2023, p. 16). This "naturalism in the Darwinian sense" (Brandom, 2023, p. 16) is still linked to Hegel's distinction between the 'theoretical and practical relationship to nature' (Hegel, 2000 [1823/1824]; cf. Berr, 2016, pp. 94–95), Kant's transcendental philosophical epistemology and metaphysics (Kant, 1959 [1781]), and Wilhelm Kamlah's concept of experiences (Kamlah, 1973). Humans are by no means at the mercy of World 1—referred to as 'nature' in the sustainability discourse—but are able to describe it, give it meaning, and not only enjoy its beauty and grandeur in the focus of a landscape aesthetic, but also deal with its arbitrariness (Hegel, 2000 [1823/1824]) and relentlessness (for more details, see Kronenberg, 2023). All these ways of dealing with nature take place in descriptions whose quality is measured not only by a standard of 'consistency with reality' but also by whether they can prove useful in solving problems.

However, since these references are context- and purpose-relative as well as contingent, this 'post-representational' understanding of human World 1 or natural conditions is associated with an increase in responsibility. The weaker the ability to justify due to the lack of a reality preceding the description, the greater the effort required to justify evaluative or normative conclusions addressed to third parties based on causal effects of World 1 (Ott, 1996). World 1, or nature manifests itself, on the one hand, in a "normative [...] authority of the represented over the representations", which takes the place of God and to which, as a causally effective "non-human authority," the "cognitive practices of humans are supposedly normatively subject" (Brandom, 2023, p. 32). According to Rorty, there is a "*responsibility* of representations towards the represented" (Brandom, 2023, p. 32). However, this responsibility exists as the "obligation to answer questions about the reasons that legitimize this authority" (Brandom, 2023, p. 33). According to Rorty, neither God nor a reality that is independent of description and normatively exemplary—i.e., the aspects of a monistic substantial reality split off from the Spinozist formula "Deus sive natura" (God or nature)—can withstand critical examination of their claim to normative authority. Therefore, as humans, we are "only responsible to each other and not bound to any authority [...] that stands outside our practices" (Brandom, 2023, p. 33).

It is these practices, i.e., negotiation processes, that determine how we respond to the causal challenges of World 1—for example, in questions of sustainable development, which should then not be guided by supposedly objective standards of nature but by standards of responsibility that humans owe to each other in the face of sustainability-related challenges and problems. World 1 has no normative authority—only a democratic, free negotiation process has that—but merely causal power with a character of necessity to which people want or need to respond—for example, in the face of repeated flooding in river valleys, building protective dikes so that no *one* is harmed. 'Nature' would still exist even if humans succeeded in destroying the Earth.

9.2 Triangulation of Worlds

In addition to the described danger of understanding World 1 or 'nature' as a given ontic instance or 'reality' and, derived from this, as a normative authority (cf. for clarification: Birnbacher, 1997, 2006), there is a danger of tending toward teleological exaggerations or utopian totalizations in the course of conceptualizing and operationalizing sustainable development. In our opinion, this danger can be effectively countered if sustainable development is understood in such a way that, in line with Karl Popper's three-world theory (see Sect. 2.1), it refers not only to the problems in World 1 (matter, physical reality), but also to World 2 (individual experience and subjective perception) and World 3 (the world of objective content, such as ideas, theories, and cultural constructs). In order to grasp these relationships, a triangulation of the three worlds is necessary to address the corresponding, sometimes complex problem areas in such a way that teleological exaggerations or utopian totalizations regarding the purposes and goals of sustainable development prove to be unsuitable.

With regard to the material dimension (World 1) of sustainable development, it is particularly important to preserve and protect natural resources, ecological balance, and biodiversity, as called for in the well-known three-pillar model of sustainability (see Grunwald & Kopfmüller, 2012; Pufé, 2017). In this context, a teleological exaggeration would mean understanding sustainable development as a clearly predetermined, linear progression leading to a final (and allegedly problem-free) state (climax). To counteract this, it is advisable to view sustainable development as a dynamic and open process that is perceived and evaluated differently on an individual (World 2) and social (different Worlds 2) level and interpreted and communicated differently on a cultural (World 3) level. From this perspective, this process must therefore be constantly adapted to the changing conditions of the three worlds, reviewed, and, if necessary, renegotiated. On the one hand, this requirement is also in line with Popper's diagnosis of the 'fallibility' and thus contingency and need for revision of all knowledge and all approaches to the world (Popper, 1963), and with his principle of 'falsificationism' (Popper, 1935), which calls for us not to strive for a definitive solution (which may be claimed to be teleologically predetermined) but to seek constantly improvable approaches that can respond to changing conditions of sustainable development by revising unsuitable knowledge or practices. This renunciation of revision-immune knowledge or corresponding practices also represents a strong bulwark against the widespread tendency toward totalization (Popper, 1945/1992).

With regard to individual experiences and subjective perceptions (world(s) 2), understanding sustainable development from the perspective of the individual plays a decisive role. A utopian view of sustainable development quickly runs the risk of viewing the individual as part of a perfectly functioning system or a desirable totalitarian society in which all human needs are harmoniously fulfilled, all problems are solved uniformly, and all challenges are mastered collectively. Here, too, it is important to recall the diagnoses and warnings of Popper, who reminds us of the disastrous link between utopian and teleological thinking and aesthetic and romantic longings for harmony and perfection—whether in interpersonal relationships or in the relationship between humans and nature—and shows how these longings for harmony and perfection can lead to oppression, totalitarianism, and terror (cf. Kühne et al., 2022; Popper, 1965, 1945/1992). In order to counteract such exaggeration, it is advisable to take into account the different cultural and social perspectives that bring with them a diversity of individual experiences and needs. It is neither sensible nor practical to develop a model that is supposedly perfect and claims to solve all problems in a golden future. Instead, different models should be developed that compete for the best ideas, practices, and proposed solutions and are put up for discussion in negotiation processes. Renouncing perfection, universal validity, totality, and guaranteed development paths may give rise to accusations that the models offered are unacceptable or inferior *because* they supposedly or actually cannot offer a perfect solution. This typical fallacy, which leads to a 'perfection trap', is based on the 'fallacy of unattainable perfection' (Edmüller & Wilhelm, 2009, p. 85). The only response to this is that—as long as there is no better alternative in sight—it is a mistake to reject a position because it is not perfect.

With regard to culture and social structures (World 3), sustainable development should be viewed as a contingent, linguistically defined cultural and social construct. A utopian totalization would necessitate that all cultures and societies develop in the same direction in order to achieve the desired ideal end goal as quickly and reliably as possible. To avoid such consequences, it is advisable to view sustainable development as a processual exchange of ideas and practices based on the recognition of the diversity of cultures, societies, and social and political systems. Instead of striving for complete standardization, it is more promising to create pluralistic, flexible models of sustainable development that are also tailored to local conditions.

9.3 Constructive Criticism

Within the framework of a neo-pragmatic understanding of the scientific process, criticism is not the end point of analysis but rather the starting point for ongoing understanding (cf. Brandom, 2001, 2008, 2011; Putnam, 1995a, 1995b; Rorty, 1979, 1997 [1989], 2023a; Shusterman, 2022). It does not aim to definitively refute or expose existing concepts but rather to further their productive development. In this sense, criticism is not a destructive act but a creative process of renewal: it should open up spaces for discussion in which existing perspectives can be examined and new contexts of meaning can be developed. Its task is to produce new metaphors—new symbolic and linguistic orders that make it possible to think differently about sustainability and life chances under changed conditions (cf. Rorty, 1989, p. 14). This idea of constructive criticism is rooted in the thinking of Richard Rorty, who emphasizes that knowledge does not aim at the discovery of objective truths but at the invention of new descriptions of the world (Rorty, 1989, p. 14). Criticism appears here as a redescription: as a conscious rewriting of what is considered real, meaningful, or just. Criticism thus becomes a form of cultural creativity that breaks down existing assumptions without replacing them with new dogmas. In this sense, criticism is understood as an ongoing process of shifting meanings that does not conclude scientific and social understanding but rather energizes it (cf. Marquard, 2007, p. 117).

For an understanding of sustainability that is oriented toward life chances, this means understanding criticism as a learning-oriented and dialogical process. Sustainability is not a closed concept but an open field of negotiation anchored in social, cultural, and ecological contexts. It articulates itself in normative tensions—between ecological resilience, social justice, economic development, and individual freedom. Constructive criticism opens up spaces for action and reflection precisely by questioning established patterns of interpretation and seeking to enable alternative interpretations. The context of Agenda 2030 and the SDGs in particular shows that such processes are necessary—for example, when the concept of quality education (SDG 4) is not only reflected in curricular terms but also as a cultural practice, power relations, and a vision for the future, or when redistribution processes (SDG 10) are read not only as technocratic control but as an expression of epistemic and political interpretive relations.

This form of criticism is always also self-criticism. Scientific integrity requires that one's own theoretical premises, methodological decisions, and normative objectives be made transparent and open to discussion. Such (self-)reflexivity is indispensable in sustainability research in particular, because its results are not only descriptive but also always action-oriented. As Stederoth (2011, p. 1354) emphasizes, every criticism requires a criterion that arises from its respective purpose. In the spirit of neopragmatism, this criterion does not lie outside the discourse but in its concrete function: theories, models, or programs must be measured by their ability to clarify real problems and open up responsible options for action (see Sect. 2.2.2).

Criticism in this sense is therefore purpose- and practice-oriented, not absolute or transcendent. It does not follow an ideal of pure objectivity, but rather an ideal of shared learning and understanding processes (cf. Putnam, 1997, p. 252; Rorty, 1989, p. 14). This contrasts with forms of criticism that judge exclusively from a position of normative superiority. The idea of an open, pluralistic practice of criticism prevents criticism itself from becoming a new dogma—and makes it a prerequisite for a sustainability discourse oriented toward life chances. For sustainability science, this means shaping criticism in such a way that it offers orientation without creating new absolutisms—in other words, reflexively, not dogmatically, experimentally.

Constructive criticism thus contributes to shaping the discourse on sustainability as an open process of understanding. It rejects the idea that scientific knowledge is fixed once and for all, and instead understands knowledge as temporary, relational, and situation-specific. This attitude is central to an understanding of sustainability that is oriented toward life chances and aims at social participation, justice, and openness to the future. Only through continued criticism can the conditions under which people are able to shape their lives in harmony with ecological and social necessities be expanded.

In this sense, constructive criticism is an integral part of sustainable practice. It is not only aimed at gaining knowledge but also at enabling new spaces for living and acting. It keeps the discourse on sustainability flexible, pluralistic, and capable of learning—and this is precisely where its life-opportunity-oriented potential lies: in its ability to open up new semantic, normative, and practical horizons that make living together under the conditions of the twenty-first century sustainable for the future.

9.4 Avoiding Pain and Humiliation

Redescriptions in the neo-pragmatic sense do not aim to destroy vocabularies, but rather to transform them in an appreciative manner (Rorty, 1989; operationalized in scientific practice, among others, at Kühne, 2024; Kühne et al., 2025). They operate beyond an epistemic mentality of 'Veni, Vidi, Vici' (Kühne, 2025): the criterion is not superiority or 'truth' (in the sense of correspondence theory) but rather translatability and the development of useful descriptions. Redescribing, on the other hand, is often associated with humiliation—even when neopragmatism seeks to avoid it: After all, as already explained, the most effective way to cause people pain is to humiliate them "by making everything that seemed particularly important to them appear futile, outdated, powerless" (Rorty, 2021 [1989], p. 153). According to Rorty (1989), humiliation can be understood as the most painful form of human experience when people lose the feeling that their interpretations of life are still meaningful. A redescription that attempts to avoid humiliation therefore seeks to connect the new with the existing, rather than staging breaks and excluding and marginalizing them. In this sense, redescription can be understood as a cultural process of developing vocabularies in which the old is not devalued but woven into new contexts (Calder, 2003; Topper, 1995). This is an attitude rooted in the contingency assumption of neopragmatism: no vocabulary is capable of claiming a privilege of truth; orders of meaning are fundamentally changeable, even if meanings can be temporarily fixed, which means drawing on the discourse theory of Laclau and Mouffe (Laclau & Mouffe, 1985), which makes Rorty's vocabulary theory more sensitive to power (see Kühne et al., 2025). However, vocabularies can only be sustainable in the long term if they are perceived as respectful by those affected. From a neopragmatic perspective, descriptions of speech are characterized by empathy; change is not carried out against those involved, but with them. In this sense, one can also speak metaphorically of a "transformation in an inhabited house" (Lohmann et al., 2024): theoretical and practical changes are made in such a way that they take everyday realities seriously and remain emotionally accessible. Social conflicts often arise not from incompatible interests, but from a lack of recognition (e.g., Berr et al., 2026a; Honneth, 1992; Kühne & Berger, 2026). People experience their perspectives as delegitimized when dominant vocabularies devalue them as irrational, backward, or unworldly. Neopragmatism formulates an alternative to the claim to "the one underlying truth" by replacing claims to truth with openness to dialogue. It replaces epistemic hierarchies with an ethic of compassion and respectful difference. In Rorty's figure of the 'liberal ironist' (Rorty, 1989), awareness of the contingency of self, language, and society is combined with the desire not to hurt or shame others, but rather to keep them in conversation. This ties in with the idea of compassion as an attitude that guides knowledge, developed by Lohmann et al. (2024): Building on Fichte's theory of emotion, compassion is understood here not as a sentimental impulse, but as a cognitive sensitivity to otherness. Compassion is understood as the ability to open up spaces for affective resonance (see also Rosa, 2016; Schlottmann & Rosa, 2023). The experiences of others are not objectified here, but rather shared. Compassion thus becomes a medium of understanding between divergent worlds. It thus represents a basic condition for a humiliation-free redescription.

Such an attitude is also of central importance for sustainable transformations. This centrality is measured not least by the significance of the challenges facing the concept of sustainable development. The example of the energy transition illustrates this particularly well (among many others: Kühne et al., 2025; Lohmann et al., 2024; Thiel & Kühne, 2025): Irresolvable conflicts often arise less from conflicting goals or factual issues than from forms of communication that cause pain. Citizens thus experience their aesthetic and, in particular, emotional connections to spaces interpreted as landscapes as devalued. This is especially true when these connections are dismissed as irrational or particularistic in the discourse on infrastructure projects. This creates epistemic humiliation, which undermines trust, blocks learning processes, and, last but not least, blocks further dialogue. A neopragmatic and life-chance-oriented redescription offers an alternative approach here: instead of addressing conflicting parties in a technocratic or moralistic way, it understands conflicts as a call for translation. This means that technical rationalities must be mediated with everyday interpretations and affective experiences. Ultimately, no single vocabulary has privileged access to the 'truth'; rather, it is important to keep people engaged in dialogue and to seek locally and regionally acceptable solutions in order to prevent large sections of those affected from adopting vocabularies that lie beyond the context of democratic negotiation processes (see, for example, Eichenauer et al., 2018; Reusswig, 2019; Reusswig et al., 2016). In this context, Lohmann et al. (2024) speak of the need to create 'spaces of resonance for compassion'. This refers to communicative and institutional arrangements in which divergent positions can be articulated without shame. In this way, transformation can become reasonable, as it connects to self-relationships and creates opportunities for respectful change. Understood in this way, the energy transition can become a laboratory for empathic redescribing: sustainable development is thus achieved not through the communication of moral superiority but through affective and communicative sensitivity. The success of such processes is measured less by CO_2 balances or expansion targets. Rather, it is measured by the ability to avoid pain and humiliation without suppressing conflicts. Such a 'humanized culture' (Rorty, 1989), in this case of sustainable development, is based on such a balance. Namely, on the combination of compassion, mutual recognition, and openness to dialogue. These are prerequisites for expanding shared opportunities in life.

9.5 Primacy of the New Metaphor Over Theory

A new metaphor can be understood as a semantic bridge that gradually expands the scope of interpretation, making systemic change appear less abrupt than a direct theoretical redescription (for more on this, see:. In pluralistic discourse arenas—such as the discourse on sustainability—a radically new theory can encounter resistance or disciplinary resistance. Metaphors, on the other hand, take up familiar figures of thought, transform them partially, and open up possibilities for association with less cognitive and affective tension (Rorty, 1989, p. 80).

From a pragmatic-neopragmatic perspective, this represents a strategic priority: language use is not merely an accompaniment but a driving force behind shifts in meaning. Richard Rorty's philosophy, for example, emphasizes that theory does not represent a superordinate layer of judicial truth but rather tools in discourse that unfold in language (Gascoigne & Bacon, 2023). In such a context, it is plausible to intervene metaphorically before offering a systematic theory. In the field of environmental and sustainability communication, the importance of metaphors is well documented empirically: Reijnierse et al. (2025) show that metaphors in environmental communication translate complex issues into familiar images, thereby increasing understanding; Flusberg & Thibodeau (2023) analyze common metaphors such as "ecological footprint" or "planet as home," which have an emotional and pictorial effect (Flusberg & Thibodeau, 2023).

In terms of an understanding of sustainability focused on life chances, this priority given to metaphors can look like this in concrete terms: instead of immediately formulating a comprehensive systems theory of life chances, metaphors such as 'opportunity ecosystem', 'flow of possibilities' or 'network of living conditions' could be introduced. They implicitly illustrate interconnectedness, dynamics, feedback loops, and structural inequalities—without having to spell out all the theoretical implications at the outset. These images can resonate in discourse, politics, and civil society and be developed in a differentiated theoretical way later on.

The metaphor-based approach offers several advantages: less resistance to new terms, greater connectivity to existing interpretive spaces, and a smooth transition to further theory formation. However, this approach must not descend into arbitrariness: metaphors must be critically reflected upon in discourse and made transparent in the course of the process so as not to obscure power structures, create interpretive hegemonies, or produce linguistic dominance (cf. Lakoff & Johnson, 2003 [1980]).

Considering that sustainability is largely conveyed through language and that many actors cling to familiar ways of thinking and expressing themselves, a metaphorical extension can often offer a more suitable introduction than a direct description. The primacy of metaphor thus opens up a gentler, more gradual possibility for transforming existing interpretive practices step by step.

9.6 Benefit Orientation

The demand to link a redescription to utility orientation responds to a central tension in normative theory formation: on the one hand, the expectation that theoretical designs will be effective in real fields of practice; on the other hand, the danger that a purely instrumental logic of efficiency will narrow what is normatively relevant. From a neo-pragmatic perspective, it is therefore plausible to reconstruct the concept of utility in such a way that it can be tested in practice, remains open to revision, and at the same time remains normatively compatible. In the pragmatic tradition—especially in the work of John Dewey and William James—beliefs and theories are not primarily seen as representations of the world, but as tools for problem solving and for guiding action. Dewey (1938) emphasizes that knowledge should be understood as part of experience and that it derives its meaning from the process of inquiry, i.e., the investigative examination of practical problems (Dewey, 1938). William James (1907) argues similarly that the truth value of an idea lies in its "cash value"—that is, in its practical consequences for life and action (James, 1907). Richard Rorty's (1979, 1989) neo-pragmatism continues this line of thinking by understanding truth and objectivity not as metaphysical quantities, but as context-sensitive results of shared language games. According to Rorty, theories derive their value not from correspondence with an external reality, but from their usefulness in the practice of mutual understanding (cf. Koopman, 2009; Rorty, 1989).

Applied to the context of sustainable development, this means that utility orientation must be understood not in technocratic terms, but as communal, learning, and socially embedded. A redescription in the neo-pragmatic sense is characterized by problem-solving ability—this does not mean maximizing efficiency, but rather the ability to open up scope for interpretation and action, to moderate conflicts, and to initiate collective learning processes (Prasad, 2021). Thus, a redefinition that reinterprets sustainability can be convincing not only through better control, but also through its connectivity to everyday practices, shared values, and participatory processes. The discussion surrounding Agenda 2030 and the SDGs in particular shows that a technocratic interpretation of utility orientation is problematic: Many objectives—such as resource efficiency or emission reduction—are operationalized in metric parameters that are often decoupled from social participation and quality of life. A neo-pragmatic approach would not reject these goals, but would translate their meaning back into everyday contexts: sustainability as a practice of shared living, in which benefits arise through understanding, learning processes, and institutional cooperation. For example, the goal of 'sustainable urban development' (SDG 11) cannot then be assessed primarily by efficiency metrics, but rather by the extent to which living spaces are created that promote democratic participation, social inclusion, and ecological resilience at the same time (cf. Agyeman, 2013).

However, the concept of utility itself is also ambivalent. Uncritical application can lead to re-technocratization—for example, when utility is again measured according to efficiency criteria or defined by expert monopolies. Rorty's (1989) warning against the "tyranny of vocabulary" is central here: any system of interpretation can become hegemonic if it absolutizes its own premises. Responsible redescription must therefore remain fallibilistic—always open to revision and dissent (cf. Putnam, 2002). Benefit orientation in the sense of neopragmatism is thus not directed toward control, but toward joint learning. In practical terms, this means that benefit-oriented redescriptions in sustainability policy perform a dual function: they combine empirical problem-solving capabilities (e.g., local climate adaptation, resource use, mobility) with normative reflexivity (justice, participation, freedom). The benefit of a redescription then lies not in increasing efficiency, but in improving collective lifestyles—in the ability to enable democratic learning processes, negotiate conflicts, and stabilize shared meanings.

It thus becomes clear that redescriptions are only useful where they prove themselves in practical contexts and can be characterized by openness to criticism and understanding. In this context, problem-solving ability means the ability to resonate, not to dominate; to communicate, not to technocracy. Thus, utility orientation is not a measure of technical efficiency, but an expression of a communally shared practice of life that is constantly being renegotiated.

9.7 Multiplicity Instead of Truth

If neopragmatism fundamentally emphasizes that, instead of direct access to reality, there can only be descriptions of this reality, and that these descriptions must be measured not by the standard of "truth" as correct correspondence with the represented reality, but by the standard of their usefulness and suitability for solving problems and mastering challenges, then it could be concluded that there is such a thing as a 'best description' or a 'most useful measure' for achieving the goals of sustainable development. Furthermore, according to the next conclusion, it would then also be necessary to reach an agreement, a consensus on this—then progress towards achieving these goals would also be possible. From a neo-pragmatic perspective, however, it can be shown why such progress in the sustainability debate does *not* arise from consensus on an 'objectively best' description or a 'definitively most useful' measure, but rather from the expansion and maintenance of a free space for discussion and argumentation, i.e., through the pluralization of arguments, positions, theories, perspectives, and vocabularies. The following variants of reasoning can be distinguished as central arguments for this assertion—bearing in mind that, in Richard Rorty's understanding, 'reasoning' does not mean 'proving' or 'substantiating'. Rorty understands 'justifying' rather in the sense of 'unfolding reasons that speak for this conviction' (Rorty, 2001a, p. 259). People *can* talk to each other, convince each other or persuade each other instead of doing violence to each other (Rorty, 1988).

9.7.1 Scientific Theoretical Justification

Neopragmatism replaces the orientation toward 'truth' as discovery and correspondence with an objective reality with a pragmatic orientation toward provisionally proven usefulness and suitability, in terms of vocabulary theory through constantly contingent and revisable redescription and recycling, and in terms of discourse theory through orientation toward a continuous and ultimately inconclusive conversation—'keeping the conversation going' (Rorty, 1979): "To see keeping a conversation going as the sufficient goal of philosophy, to understand wisdom as the ability to contribute to a conversation, means to see human beings not as beings that one hopes to be able to describe accurately at some point, but as the producers of descriptions" (Rorty, 1981 [1979], p. 409). In this sense, Rorty even understands conversation as "the highest good of discursive creatures" (Rorty, 2001b, p. 38) and the expansion of conversation—in the sense of an "expansion of the we" (Rorty, 1988, 2000, 2023a)—in which new descriptions become possible, as a meaningful orientation. Such an expansion leads to the advocacy of a pluralistic discourse landscape, which makes it possible in the first place for alternative descriptions of ecological, social, cultural, political, economic, or other contexts to become visible and negotiable. Clinging to a single "best" or 'most useful' description would—similar to clinging to a search for definitive 'truth'—hinder or even end the process of collective and individual learning, because it would become entangled in counterproductive and ultimately undecidable disputes about the 'true' or 'correct' description or consideration of usefulness. But then what Popper calls 'error elimination' (Popper, 1935) in the context of his falsificationism and Rorty calls 'conversation rather than confrontation' (Bernstein, 1983) would become impossible. In the context of sustainability in particular, innovation and thus progress are often the result of unexpected shifts in perspective: 'reasonable plurality' (Rawls, 1993; cf. Reese-Schäfer, 2023) creates argumentative friction between different points of view, which can lead to new and creative solutions to problems. A consensus on the 'most useful measure' would prematurely end the discourse, whereas the unbroken diversity of possible points of view keeps the discussion or negotiation process open and can promote adaptability to diverse and often changing challenges.

9.7.2 Antiteleological Reasoning

The rejection of teleological narratives and historical processes, as already explained, logically leads to the rejection of any myth of progress (Popper, 1965). In other words, neopragmatism replaces the ideal of a final truth or a final state of the world with the ideal of continuous improvement. From this perspective, progress is understood not as a final state, but as an ongoing movement of dialogue. Sustainable development cannot therefore be determined by a master plan based on anticipated end states of historical processes or specific developments, but only through open-ended negotiations—it always remains a "work in progress" based on "small steps" are only provisionally valid and proven. Teleological or utopian totalizations also contradict this attitude because they would negate the processual nature of learning in discourses and negotiations.

9.7.3 Ethical-Political Justification

Discourses or negotiations require a certain degree of common ground among the participants, at least formal rules of procedure that are accepted by all (Dahrendorf, 1972; Rawls, 1975 [English original 1971]). Rorty does not seek to base such common ground on an 'objective' return to 'reason', 'objectivity' or 'nature', but rather on solidarity (Rorty, 1988) and thus on a practical way of living together. In comparison to 'classical' liberal thinkers such as Rawls, who still strive for a universally valid justification of a liberal-democratic community, he considers an 'extension of the we' (Rorty, 1988, 2000, 2023a; cf. also Sect. 3.4) as the reference point for his considerations. From this perspective, progress does not come about by proving that one position is 'truer' or 'more correct' or 'more useful', but through an increase in understanding, empathy, mutual responsiveness and thus responsibility. Neopragmatically, the concept of solidarity thus refers to an ethics of responsibility that contrasts with postulates and moralizations based on ethics of conviction, which are oriented toward an 'objective' and 'true' nature or reality or 'right' and 'universal' moral principles and therefore lead to the exclusion of those who think differently and thus to a reduction of the 'we' in Rorty's sense. Sustainability in the neo-pragmatic sense is therefore to be understood and conceived as a practice of shared responsibility, not as a system of universal truths and moral certainties. This rejection of an ethical orientation toward (absolute) truths and certainties also means that, in the neo-pragmatic sense, freedom cannot be secured through consensus, but rather, conversely and more pointedly, only through the right to dissent. For if different communities are allowed to articulate their own ideas of the good life, progress, sustainability, or nature, then moral progress arises not through standardization, but through the recognition of difference. As Rorty emphasizes, emancipation means expanding the circle of conversation, not closing it. In the neo-pragmatic understanding, progress in sustainable development does not come about by finding the *right description*, but by *continuing to talk together* (Rorty, 1981 [1979], pp. 421–427)—in freedom and without the pressure to agree. Sustainability is therefore not a fixed goal, but rather the subject of a conversation that must be kept open.

9.8 Mutual Reference Between Self-creation and Solidarity

A neo-pragmatic further development of the concept of sustainability therefore begins where classical understandings reach their limits. Classical understandings often assume that sustainable development is a system of goals to be balanced between economics, ecology, and social issues, between culture and politics. From a neo-pragmatic perspective, sustainable development instead appears as a dynamic structure of social meaning production. This is based on the dual condition of human freedom: firstly, the ability for private self-creation and, secondly, the necessity of public solidarity. These two dimensions are not opposites. Rather, they can be understood as coordinates of a resonance space in which individuals and societies generate their orientations. With Rorty (1989), self-creation can be understood as a contingency-conscious, creative rewriting of the self. This private dimension is a source of diversity and innovation. It keeps the sustainability discourse open to new forms of what can be considered a 'good' or 'sustainable' life. However, this private self-creation depends on a culture in which cruelty is avoided and differences are respected. It is therefore structurally dependent on a society that guarantees solidarity.

At this point, we can refer to Ralf Dahrendorf's concept of life chances (Dahrendorf, 1979). Freedom arises in the tension between options (possibilities for action) and ligatures (connections and meanings). A sustainable society does not develop where these poles are balanced or where one is subordinated to the other. Rather, it develops where options and ligatures can enter into productive resonance. Options open up spaces for self-creation, while ligatures create orientation and belonging.

In this sense, the process of sustainable development must create conditions under which people are free to choose what they want to commit to voluntarily. This results in a relationship that supports both individuality and community. This perspective leads to a conceptual expansion of the concept of sustainability:

1. Sustainability is not sought as a stable end state. Rather, sustainable development is understood as a process of reflexive and affective resonance between individual life plans (private self-creation and the pursuit of options) and collective bonds (ligatures and solidarity).
2. Sustainable development is understood as a cultural practice that institutionally anchors learning ability, translation work, and respect for different worldviews and self-relationships.
3. The goal is not balance or equality, but acceptibility and creativity: transformations should enable further self-creation while securing and expanding social cohesion.

In this understanding, compassion (see Sect. 9.4) takes on a knowledge-guiding attitude that is introduced in social science sustainability research. Referring to Fichte, we understand compassion not as a sentimental reaction, but as a cognitive sensitivity to otherness—as the ability to open up spaces of affective resonance in which the other can be experienced in its own logic. This embeds individual options morally and shapes voluntary ligatures in such a way that they do not bind normatively, but rather enable resonance. Sustainability conflicts often escalate where this resonance is disrupted (Kühne & Berger, 2026). This occurs when options are experienced without recognition or when ligatures restrict freedom without any associated meaning. A life-chance-oriented redescription therefore understands sustainable development as the ongoing negotiation of the relationship between individual freedom and collective responsibility. Ultimately, this is an open, fallibilistic process of social meaning-making and perpetuation. With Martin Müller (2014), it can be pointed out that Rorty's figure of the liberal ironist symbolizes the ethical model of this attitude: The liberal ironist is aware of the contingency of her own convictions and also shows solidarity with those whose life plans have taken a less fortunate turn. After all, contingency means that not every decision proves successful and that the social context can sometimes have a restrictive effect. Her irony is not cynicism. Rather, it is an expression of responsible individualism that co-constitutes self-creation and public concern. In this sense, sustainable development succeeds where people can revise their own descriptions of the good without destroying the common framework of understanding. Sustainable development should therefore not be understood as a technical control concept or moral imperative, but as a practice of creating resonance. This intertwines individual creativity, compassion, and solidarity. Options and voluntarily chosen ligatures thus form, in mutual intertwining, the foundation of an open, adaptive, and humane culture of sustainable development.

9.9 Sensitivity to Conflict Theory

Since conflicts are inevitable in the process of sustainable development, the question arises as to how these should be structured and regulated in order to fulfill their purpose productively. Conflicts can also be understood in a neo-pragmatic way as an expression of competing vocabularies (Rorty, 1989), because vocabularies shape "mutually distrustful (and *not*: incomprehensible) communities of justification" (Rorty, 2023a, p. 178). Ideally, these communities can nevertheless remain in dialogue with one another and mutually recognize the legitimacy of each other's vocabularies. However, this is only possible if these conflicts over vocabularies are understood and handled as a functional struggle for suitable solutions and not as a battleground for resentment-driven moralizing or destructive attitudes and values (Berr & Kühne, 2019, 2024; Gehlen, 2016; Grau, 2017; Kühne, 2019b; Lübbe, 2019; Luft, 2019; Neuhäuser & Seidel, 2020), which strive for solutions with absolute claims to validity.

In all societies, conflicts arise time and again due to differing interests, needs, and demands, and this occurs in almost all areas of life (cf. Berr, 2019; Dahrendorf, 1972; Thiel & Kühne, 2024), for example in rural areas (Berr & Jenal, 2019; Berr et al., 2024a, 2024b; Kühne, 2018; Kühne & Berr, 2023). Ralf Dahrendorf has shown that conflicts should not be understood as disturbances, but rather as inevitable and therefore common occurrences (Dahrendorf, 1972). They are a structural feature of human communication and interaction and therefore by no means an evil to be avoided—not even in conflicts over landscapes (Kühne, 2019a), for example in the context of sustainable development. Scientific conflicts over vocabulary can also be described as 'dissent' (Hubig, 2007). These can never be completely avoided, nor is it fundamentally sensible or expedient to try to overcome or resolve dissent (cf. for details: Weber et al., 2018), because dissent has both the "potential for subjective identity formation and free choice of action" (Hubig, 2007, p. 150) against the epistemic or moral tendency of consensus to become fixed (cf. Hubig, 2007). In any case, expectations of consensus are often too high, as the necessary "talking to each other" is deeply "ambivalent" (Ottmann, 2012, p. 118): In the traditional sense of Aristotle (2009), it can lead to cooperative "community," but in the sense of Hobbes (2017 [1642]), it can also lead to ongoing conflict and dissent and even enmity (cf. Grau, 2017; Mouffe, 2007, 2014; Schmitt, 1933, 2011 [1967]) and, in extreme cases, to civil war. Excessive expectations of consensus can lead to a 'compulsion to achieve consensus' (Berr, 2019) and to 'ethical rigorism' (Apel, 1973; Gethmann & Sander, 2004, p. 115).

Insofar as certain framework conditions can be adhered to when dealing with conflicts, Dahrendorf attributes social productivity to them, as they can contribute to social progress (Dahrendorf, 1957, 1969, 1972, 1992; Kühne, 2017, 2018, 2019a, 2020; Niedenzu, 2001). According to Dahrendorf, conflicts should even be understood as a basic prerequisite for human coexistence and as a driving force for social change: "Conflict is therefore the creative core of all society and the opportunity for freedom" (Dahrendorf, 1961, p. 235). In his conflict theory, Dahrendorf identified the conditions under which a conflict can be creative (in particular in Dahrendorf, 1972). In the following, we will apply these conditions in general to the conflict between different vocabularies, as is the case, for example, in conflicts in the context of sustainable development.

The most important condition is that conflicts should always lead to constructive dialogue, meaning that vocabulary should not be used merely as a medium for representing interests, exercising power, self-promotion, or destructive intentions. For this to be possible, a few other conditions must be met. Firstly, this includes the fundamental recognition of the legitimacy of conflicting interests, intentions, and values; with regard to vocabularies, this means recognizing their logic, purpose, and linguistic characteristics. Secondly, the focus is on dealing with the specific conflict (not general principles); in the conflict over vocabularies, the question is therefore the suitability of specific terms, methods, or theories with regard to specific purposes and goals (not an entire vocabulary or an entire scientific direction or discipline). Thirdly, reliable negotiating partners are needed who can speak legitimately for their side; in the conflict over vocabularies, there is a corresponding need for competent speakers who have an in-depth and targeted command of their vocabularies and do not want to strategically dazzle with half-knowledge. Fourthly, it is necessary to adhere to democratic and communicative rules of the game that do not favor any of the parties or opponents in the conflict and must therefore be regarded as equal. In terms of vocabulary theory, this means recognizing vocabularies as equal, provided they meet relevant scientific standards, since every vocabulary is part of a "linguistic" (Putnam, 1990, p. 43) or scientific division of labor (Tetens, 1999) can explain specific aspects and is therefore indispensable—especially in view of the complex challenges of sustainable development.

Insofar as Dahrendorf's conflict theory does not necessarily aim at consensus or compromise, nor at the suppression of opposing positions, conflicts between vocabularies should not aim to level their differences in a 'super-theory' or 'super-paradigm', to defuse them in a factual or postulated scientific mainstream, or to silence them through discourse exclusion, discrediting, or moralizing. Instead, dealing with conflicts productively requires the parties to the conflict to assess their values and beliefs as contingent and thus open to criticism, to view their own beliefs as relative in their claim to validity, and to recognize the other actors and their positions as equal. Accordingly, 'dissent management' (Hubig, 2007, pp. 147–163) does not attempt, as discourse ethics does, for example, to "define the participants by achieving or even demanding object-level consensus through acceptance of the consequences of general compliance [...]" (Hubig, 2007, p. 150). With an object-level consensus or compromise, everyone can save face, but the problems are "not solved, but merely postponed"—for example, in the case of "energy mix, transport mix, sustainable tourism, etc." (Hubig, 2007, p. 159). Instead, a *meta-level* consensus is needed on the general goal of sustainable development—avoiding unnecessary suffering, unhappiness, and cruelty (Müller, 2023; Rorty, 1989; Shklar, 1985, 1990, 2020)—and on the general procedural rules for conflict resolution and dissent management—as outlined above.

9.10 Integration and Reusability

From a life chances-oriented neopragmatic perspective, sustainable development is primarily an open learning process. In this sense, it requires not only technical knowledge or economic considerations, but also the ability to jointly generate new descriptions of societal futures and relate them to one another. The neo-pragmatic perspective underlying this book makes it clear that vocabularies must always remain provisional: they do not describe teleological target states, they are not immune to new metaphors, but rather they are offers for continued conversation (Rorty, 1989). The value of a description therefore lies not in its 'truth content' but in its reusability. The value of neopragmatic vocabularies lies in their ability to incorporate other descriptions, transform them, and use them productively in new situations. This is crucial for the project of sustainable development, as social challenges change, lines of conflict shift, and new actors enter the communication process. A vocabulary intended to describe sustainable development must therefore be inclusive, adaptable, and open to the future. Neopragmatic redescriptions can be effective precisely because they recycle old descriptions and weave them into new contexts of meaning (Kühne & Berr, 2026; Topper, 1995). New descriptions and metaphors tie in with old ones. If the goal is not to replace old vocabulary with new vocabulary (but rather to weave them together), then there is no need to immunize old vocabulary against redescribing efforts. On the contrary: they should be designed in such a way that they offer starting points for new metaphors and are open to redescriptions (among many: Kühne, 2024; Kühne et al., 2025, 2026a). Sustainable development can benefit from such a perspective: in this sense, there is no need for conceptual breaks, but rather transitional concepts that make change reasonable and do not humiliate existing patterns of interpretation and descriptions of the world, but rather represent an invitation to transformation. The concepts introduced in this chapter—compassion, acceptibility, recognition, resonance, options, and ligatures (as well as the figure of the liberal ironist)—can be understood not only as aspects of such a development, but also as tools for expanding the 'we':

- Compassion creates spaces for affective resonance. New descriptions can be absorbed in these spaces.
- Acceptibility describes the normative framework in which transformation remains possible without humiliation.
- Options and ligatures enable both self-creation and collective orientation. They thus represent mutually dependent basic conditions for sustainable development.
- Resonance describes a relational structure in which these elements can interact effectively.

Following this line of thought, reusability is not a methodological side issue or a theoretical finger game, but a criterion of sustainable development. Sustainable vocabulary based on a neopragmatic understanding therefore has the following criteria:

- It remains understandable by linking to existing references to the living world.

- They have a transformative effect in this way by opening up new perspectives.
- They are geared toward inclusivity by not excluding others, but rather creating resonance.
- They remain open to further development by enabling future redescriptions.

Following this, sustainable development becomes an open, ongoing negotiation process in which each new vocabulary is consciously designed not to exclude, but to open up, to be open to new metaphors and redescriptions. This also includes the considerations developed in this chapter. They do not mark end points, but rather intermediate stages: they represent contributions that can be taken up, varied, and woven into the process of redescription in the future. Integration thus does not mean fitting into a canon, but rather providing a resonant vocabulary. This supports people in developing their own descriptions of the future and synchronizing them with each other. Reusability thus becomes the core principle of sustainable development—not only in a material context: every new metaphor, every new concept is understood as a finite contribution to a conversation that can only be sustainable if it does not end but is continued and integrates more and more people.

9.11 Appreciation of Locally Embedded Practices

In the previous sections, it has already been suggested in a context-related manner that vocabularies, due to their purpose-oriented nature and the general scientific and linguistic division of labor, always have only a limited scope and can claim only partial validity. Universalist claims to validity and global impact within the framework of supertheories or superparadigms have always proven, and continue to prove, to be exaggerated and unrealistic upon closer inspection. This also applies to the theoretical modeling (conceptualization) and practical implementation (operationalization) of sustainable development. These difficulties are well known in discussions about the differences between 'universality vs. particularity' or 'global vs. local'. In conflict theories, this problem is reflected in the difference between general principles or solutions ('evil capitalism', 'world revolution') and concrete issues and regulations (a company and its local environmental balance, a correspondingly concrete and effective regulation); in scientific disagreements between absolute claims to validity (only essentialism, or only discourse theory, or only actor-network theory, etc. are the only correct theory) and purpose-related contingent claims to validity (for this scientific problem, this theoretical approach 'X' or this method 'Y' is epistemically useful and practicable for the time being).

In discourse and vocabulary theory, the pursuit of universality—as already shown—is reflected in the pursuit of consensus in conflicts and disagreements. This becomes abundantly clear when this universalism is not only directed at factual issues, but, in view of the inevitable failure to universalize these factual issues, takes refuge in 'moral universalism' (Kühne, 2020; Thiel & Kühne, 2025), which assumes that conflicts or dissent can be smoothed over by equating the morally good with the ethical and evaluative good (for this distinction, see Berr, 2020; Hofmann-Riedinger, 2011; Kant, 1983 [1793]) and claiming that it is *equally valid* for all people. However, this equation and indifference misses the mark in a pluralistic reality of different values and worldviews: what is "morally (good for everyone)" does not automatically coincide with what is ethical (good for us)". This ignores the diversity of people and their ways of life and communities in the form of different everyday worlds, morals, beliefs, or value orientations, which cannot be reduced to a common "lowest common denominator" (Hubig, n.d., n.p.). As a result, the same validity or applicability of a universalistic morality can lead to indifference or apathy towards the diversity of individual people. The needs and interests of individuals are then merely "material to be processed" (Kuhlmann, 1985, p. 246) or a "duty" (cf. Hoffmann, 2009, p. 319) or they are utilitarianly "offset" against each other (Höffe, 2008; Rawls, 2017 [English original 1971], pp. 211–220). Ultimately, a global 'universalism of law and morality' can then be played off against the particularity of small-scale (local, communal, regional, national) forms of morality, ethics, law, and traditions, and be upgraded as decisive in relation to them.

If these considerations can be accepted, then corresponding conclusions for spatial contexts are also plausible. In that case, the primary question is not whether and, above all, how the concepts and operationalizations of sustainable development can claim universal validity and be implemented uniformly in all spaces, landscapes, and regions of the world. Instead, it becomes clear that there is not and cannot be a single universally valid vocabulary of sustainable development, but rather a multitude of specific vocabularies that must respond to the particularities of different spaces, landscapes, and regions—such as local cultures, social customs, political conditions, religious beliefs, moral values, etc.—in order to achieve the desired effect at all. This renunciation of universality in provisional solutions is also reflected in the preference for contingent, local, and experimental practices, which Rorty understands as "improvisational" and "culture-dependent" (Rorty, 1981 [1979], 1989, 1991b, 1998).

9.12 Procedural and Substantive Justice

Sustainable development focused on life chances requires that social transformation processes enable fair procedures on the one hand and expand the range of options available to individuals on the other. Procedural and opportunity equality form a common basis for this: only when people are involved in processes in which they are respected rather than humiliated can they take advantage of the opportunities that are structurally open to them. This also means that participation can only be more than a formal gesture where real options exist. From a neopragmatic perspective, procedural justice thus means more than institutional transparency or equal participation rights. Procedures become fair when they allow for resonance, when participants are taken seriously in their experiences and concepts. But also when processes are designed in such a way that change is not experienced as an imposition, but as a reasonable translation. This gives the ethics of compassion a central role, in which compassion is understood as the cognitive sensitivity that enables us to empathize with the perspective of others without appropriating it (Berr et al., 2026b; Lohmann et al., 2024). In this sense, fair procedures are based not only on institutional rules, but also on affective openness, which creates resonance between divergent worlds. The principle of equal opportunity forms the structural basis for this procedural dimension: people need not only formal equality in procedures to shape their lives, but also actual opportunities to do so. Sustainable development must expand options and at the same time enable free choice of ligatures. The expansion of available options, whether social, economic, spatial, or cultural, is a prerequisite for people to be able to shape their own lives in times of change. Voluntarily entered into ligatures enable embedding in sustainable social contexts. In this sense, equal opportunity does not mean the equal distribution of resources, but rather the securing of life chances that both enable self-creation and strengthen social cohesion. Procedural and equal opportunity together form a resonant enabling structure (Kühne et al., 2026b). However, procedures that operate without compassion remain formalistic and create distance, which makes it easier to break off the conversation. Opportunities that are not secured by fair procedures often remain unused because they lack reliability. Sustainable development thus has a dual normative-ethical basis: on the one hand, procedures are needed that are appreciative, sensitive, and revisable. On the other hand, it must also ensure structural conditions that enable people to develop their options in times of change without losing their resonance spaces. Sustainable development succeeds where processes do not hurt but recognize. It succeeds where opportunities are not abstract but actually usable. It also succeeds where compassion is effective as the affective basis of that social resonance. Procedural and opportunity justice thus jointly constitute an order that combines self-creation and solidarity. An order that does not remain abstract when it is permeated with compassion between people.

9.13 Summary of the Redescribing of the Life-Chance-Oriented Understanding of Sustainability Based on Neopragmatism

As this chapter has made clear, the redefinition of sustainable development in terms of a life-chance-oriented neo-pragmatic approach provides neither a new model nor an alternative normative architecture. Instead, it clarifies the conditions under which sustainable development can succeed as an open socio-cultural practice. The starting point here is a departure from representationalism. For neither can the world be represented linguistically, nor is sustainability an essential principle that is universally and unambiguously defined in normative terms. Meaning does not arise from mirroring the world or trusting in the discovery of an essential nature, but rather from the use of vocabularies. This use takes advantage of the diversity of perspectives as an epistemic and social starting point. In this understanding, different forms of experience, knowledge bases, and descriptions do not constitute competing claims to truth. Rather, they are each their own approaches to interpreting and ordering complex challenges. From this perspective, conflicts are not understood as a disturbance of a desired state of harmony, but are treated as productive differences. Constructive criticism, as we have developed it in this book, means understanding alternative descriptions as an opportunity for learning and resonant further development. The central prerequisite for conversations open to transformation is the norm of avoiding pain and humiliation. This avoidance means not violating self-relationships and not delegitimizing existing life plans.

By emphasizing the primacy of metaphor over theory, we aim to generate new perspectives on sustainable development not through abstract models, but by finding new, connectable ways of description. These support the reorganization of experiences and enable social understanding. This makes it clear that sustainable development—in a neo-pragmatic and life-chance-oriented sense—cannot be determined by ultimate justifications. It can enter into a kind of 'fluid stability' when it becomes practically significant for different groups, situations, and challenges. This practical relevance also results from the mutual reference between private self-creation and public solidarity. In this context, a neo-pragmatic and life-chance-oriented sustainable development cannot privilege either exclusively individual self-designs or exclusively collective bonds. Only with a resonant relationship between options and ligatures can life chances be created. The growth of life chances for more and more people is necessary for a reasonable and just transformation. Change succeeds where local knowledge is taken into account, existing practices are respected, and new vocabularies are created in a respectful manner. This means that time-honored descriptions are not replaced, but are incorporated into new ones.

The understanding of life-opportunity-oriented and neo-pragmatic sustainable development developed here combines procedural and opportunity justice on the basis of compassion. Sustainable development in this sense requires procedures that are fair, inclusive, and resonant. At the same time, it requires structural conditions that also open up opportunities for people to participate in these procedures. Compassion acts as an affective foundation. It creates social resonance and makes it possible to acknowledge differences without having to level them out. Sustainable development thus appears as an ongoing process of linguistic, social, and institutional coordination. It does not strive for truth, but for connectivity; not for harmony as a conflict-free state, but for resonance; not for target images, but for reasonable opportunities to remain in constant dialogue.

References

Agyeman, J. (2013). *Introducing Just Sustainabilities: Policy, Planning, and Practice*. Zed Books.
Apel, K.-O. (1973). *Transformation der Philosophie: Band 2. Das Apriori der Kommunikationsgemeinschaft. Transformation der Philosophie: Vol. 2*. Suhrkamp.

Aristoteles. (2009). *Politik: Neuausgabe.* Nach der Übersetzung von Franz Susemihl mit Einleitung. In *Bibliographie und zusätzlichen Anmerkungen von Wolfgang Kullmann.* Rowohlt.
Bernstein, R. J. (1983). *Beyond Objectivism and Relativism: Science, Hermeneutics, and Praxis.* University of Pennsylvania Press.
Berr, K. (2016). Stadt und Land(schaft): Ein erweiterter Blick mit dem „zweyten Auge" auf ein fragwürdig gewordenes Verhältnis. In K. Berr & H. Friesen (Eds.), *Stadt und Land: Zwischen Status quo und utopischem Ideal* (pp. 75–117). Mentis.
Berr, K. (2019). Konflikt und Ethik. In K. Berr & C. Jenal (Eds.), *Landschaftskonflikte* (pp. 109–129). Springer VS.
Berr, K. (2020). Vom Wahren, Schönen und Guten: Philosophische Zugänge zu Landschaftsprozessen. In R. Duttmann, O. Kühne, & F. Weber (Eds.), *Landschaft als Prozess* (pp. 83–117). Springer VS.
Berr, K., & Jenal, C. (Eds.). (2019). *Landschaftskonflikte.* Springer VS.
Berr, K., Jenal, C., & Kindler, H. (2024). Landschaftskonflikte. In O. Kühne, F. Weber, K. Berr, & C. Jenal (Eds.), *Handbuch Landschaft* (2nd ed., 663–679). Springer VS.
Berr, K., Koegst, L., & Kühne, O. (Eds.). (2024). *Landscape Conflicts.* Springer VS. https://doi.org/10.1007/978-3-658-43352-9
Berr, K., & Kühne, O. (2019). Werte und Werthaltungen in Landschaftskonflikten. In K. Berr & C. Jenal (Eds.), *Landschaftskonflikte* (pp. 65–88). Springer VS.
Berr, K., & Kühne, O. (2024). Moral und Ethik von Landschaft. In O. Kühne, F. Weber, K. Berr, & C. Jenal (Eds.), *Handbuch Landschaft* (2nd ed., 647–662). Springer VS.
Berr, K., Kühne, O., & Jenal, C. (2026a). Vergangene Zukunftsräume: Eine Annäherung an Redeskriptionen und Vokabularen der Anerkennung zu Altindustrielandschaften. In O. Kühne, K. Berr, & C. Jenal (Eds.), *Neopragmatismus und Raum* (in diesem Band). Springer VS.
Berr, K., Kühne, O., Lohmann, P., & Schuster, K. (2026b). Mitgefühl und Raum: Neopragmatistische Perspektiven auf eine aktualisierte humanistische Geographie. In O. Kühne, K. Berr, & C. Jenal (Eds.), *Neopragmatismus und Raum* (in diesem Band). Springer VS.
Birnbacher, D. (1997). „Natur" als Maßstab menschlichen Handelns. In D. Birnbacher (Ed.), *Universal-Bibliothek. Ökophilosophie* (pp. 217–241). Reclam.
Birnbacher, D. (2006). *Natürlichkeit.* de Gruyter. https://doi.org/10.1515/9783110193695
Brandom, R. B. (2001). *Making It Explicit: Reasoning, Representing, and Discursive Commitment* (4. print). Harvard University Press.
Brandom, R. B. (2008). Towards an Analytical Pragmatism. *Philosophical Topics, 36*(2), 1–27.
Brandom, R. B. (2011). Vocabularies of Pragmatism: Synthesizing Naturalism and Historicism. In R. B. Brandom (Ed.), *Perspectives on Pragmatism: Classical, Recent, and Contemporary* (pp. 116–157). Harvard University Press. https://doi.org/10.4159/9780674270572-006
Brandom, R. B. (2023). Vollendung der Aufklärung: Vorwort von *Robert B. Brandom.* In E. Mendieta (Ed.), *Pragmatismus als Antiautoritarismus* (pp. 7–41). Suhrkamp.
Buschmeier, M. (2023). Sprachphilosophie: Sprachspielpragmatismus und -historismus. In M. Müller (Ed.), *Handbuch Richard Rorty* (pp. 597–614). Springer Fachmedien.
Calder, G. (2003). *Rorty: And Redescription.* Weidenfeld & Nicolson.
Dahrendorf, R. (1957). *Soziale Klassen und Klassenkonflikt in der industriellen Gesellschaft.* Enke.
Dahrendorf, R. (1961). *Gesellschaft und Freiheit: Zur soziologischen Analyse der Gegenwart.* Piper.
Dahrendorf, R. (1969). Sozialer Konflikt. In W. Bernsdorf (Ed.), *Wörterbuch der Soziologie* (pp. 1006–1009). Ferdinand Enke Verlag.
Dahrendorf, R. (1972). *Konflikt und Freiheit: Auf dem Weg zur Dienstklassengesellschaft.* Piper.
Dahrendorf, R. (1979). *Lebenschancen: Anläufe zur sozialen und politischen Theorie. Suhrkamp-Taschenbuch: Vol. 559.* Suhrkamp.
Dahrendorf, R. (1992). *Der moderne soziale Konflikt: Essay zur Politik der Freiheit.* Deutsche Verlags-Anstalt DVA.
Davidson, D. (2004). *Subjektiv, intersubjektiv, objektiv.* Suhrkamp.
de Spinoza, B. (2010 [1677]). *Die Ethik nach geometrischer Methode dargestellt: Sämtliche Werke.* Meiner.

Descartes, R. (2008 [1637]). *Philosophische Bibliothek: Vol. 597. Meditationes de prima philosophia: Lateinisch – Deutsch* (C. Wohlers, Ed.). Meiner.

Dewey, J. (1938). *Logic: The Theory of Inquiry*. Henry Holt and Company.

Edmüller, A., & Wilhelm, T. (2009). *Manipulationstechniken*. Haufe Lexware.

Eichenauer, E., Reusswig, F., Meyer-Ohlendorf, L., & Lass, W. (2018). Bürgerinitiativen gegen Windkraftanlagen und der Aufschwung rechtspopulistischer Bewegungen. In O. Kühne & F. Weber (Eds.), *Bausteine der Energiewende* (pp. 633–651). Springer VS. https://doi.org/10.1007/978-3-658-19509-0_32

Flusberg, S. J., & Thibodeau, P. H. (2023). Why Is Mother Earth on Life Support? Metaphors in Environmental Discourse. *Topics in Cognitive Science, 15*(3), 522–545. https://doi.org/10.1111/tops.12651

Gascoigne, N., & Bacon, M. (2023). Taking Rorty Seriously: Pragmatism, Metaphilosophy, and Truth. *Inquiry, 66*(10), 1773–1792. https://doi.org/10.1080/0020174X.2020.1820375

Gehlen, A. (2016). *Moral und Hypermoral: Eine pluralistische Ethik*. Klostermann.

Gethmann, C. F., & Sander, T. (2004). Rechtfertigungsdiskurse. In H. Friesen & K. Berr (Eds.), *Angewandte Ethik im Spannungsfeld von Begründung und Anwendung* (pp. 111–158). Peter Lang.

Gloy, K. (2004). *Wahrheitstheorien: Eine Einführung*. Francke.

Grau, A. (2017). *Hypermoral: Die neue Lust an der Empörung* (2nd ed.). Claudius.

Grunwald, A., & Kopfmüller, J. (2012). *Nachhaltigkeit: Eine Einführung* (2., aktualisierte Auflage). *Campus »Studium«*. Campus.

Hegel, G. W. F. (1970 [1830]). *Enzyklopädie der philosophischen Wissenschaften im Grundrisse 1830: Erster Teil: Die Wissenschaft der Logik. Mit mündlichen Zusätzen* (E. Moldenhauer, & K. M. Michel, Eds.). Suhrkamp.

Hegel, G. W. F. (2000 [1823/24]). *Hegeliana: Vol. 12. Vorlesung über Naturphilosophie Berlin 1823/24: Nachschrift von K.G.J. v. Griesheim* (G. Marmasse, Ed.). Peter Lang.

Heidegger, M. (1992). *Der Ursprung des Kunstwerkes*. Reclam.

Heindl, A. (2023). Richard Rortys Kritik der Erkenntnistheorie: Demokratischer Anti-Repräsentationalismus. In M. Müller (Ed.), *Handbuch Richard Rorty* (pp. 581–596). Springer Fachmedien.

Hobbes, T. (2017 [1642]). *De cive/Vom Bürger: Lateinisch/Deutsch*. Reclam.

Hoffmann, T. S. (2009). *Wirtschaftsphilosophie: Ansätze und Perspektiven von der Antike bis heute*. Marix.

Hofmann-Riedinger, M. (2011). gut/das Gute/das Böse. In M. Düwell, C. Hübenthal, & M. H. Werner (Eds.), *Handbuch Ethik* (3., aktualisierte Auflage, pp. 387–391). J.B. Metzler.

Honneth, A. (1992). *Kampf um Anerkennung: Zur moralischen Grammatik sozialer Konflikte*. Suhrkamp.

Höffe, O. (Ed.). (2008). *Einführung in die utilitaristische Ethik: Klassische und zeitgenössische Texte* (4., überarbeitete und erweiterte Auflage). Francke.

Hubig, C. (o.J.). *Interview: „Es geht um einen echten Ausgleich für Schäden"*. https://www.cssa-wiesbaden.de/nachhaltigkeitsinitiativechemie/nachhaltigkeit-dissensmanagement/

Hubig, C. (2007). *Die Kunst des Möglichen II: Ethik der Technik als provisorische Moral. Edition panta rei: Bd. 2*. transcript.

James, W. (1907). *Pragmatism: An Old Way for Some New Ways of Thinking*. Longmans, Green and Co.

Janich, P. (1996). *Was ist Wahrheit? Eine philosophische Einführung* (Original-Ausgabe). *Beck'sche Reihe C. H. Beck Wissen: Vol. 2052*. Beck.

Kamlah, W. (1973). *Philosophische Anthropologie: Sprachkritische Grundlegung und Ethik. BI-Hochschultaschenbücher: Vol. 238*. Bibliographisches Institut.

Kant, I. (1959 [1781]). *Kritik der reinen Vernunft*. Felix Meiner Verlag.

Kant, I. (1983 [1793]). Über den Gemeinspruch: Das mag in der Theorie richtig sein, taugt aber nicht für die Praxis. In W. Weischedel (Ed.), *Kant: Werke. Band 9* (pp. 125–172). WBG.

Koopman, C. (2009). *Pragmatism as Transition: Historicity and Hope in James, Dewey, and Rorty*. Columbia University Press. http://gbv.eblib.com/patron/FullRecord.aspx?p=4454800

Kronenberg, B. (2023). Ontologie: Taktischer Physikalismus und Naturalismus. In M. Müller (Ed.), *Handbuch Richard Rorty* (pp. 655–671). Springer Fachmedien.

Kuhlmann,W. (1985). *Reflexive Letztbegründung: Untersuchungen zur Transzendentalpragmatik*. Alber.

Kühne, O. (2017). *Zur Aktualität von Ralf Dahrendorf: Einführung in sein Werk. Aktuelle und klassische Sozial- und Kulturwissenschaftler|innen*. Springer VS.

Kühne, O. (2018). Neue Landschaftskonflikte' – Überlegungen zu den physischen Manifestationen der Energiewende auf der Grundlage der Konflikttheorie Ralf Dahrendorfs. In O. Kühne & F. Weber (Eds.), *Bausteine der Energiewende* (pp. 163–186). Springer VS.

Kühne, O. (2019a). Die Produktivität von Landschaftskonflikten – Möglichkeiten und Grenzen auf Grundlage der Konflikttheorie Ralf Dahrendorfs. In K. Berr & C. Jenal (Eds.), *Landschaftskonflikte* (pp. 37–49). Springer VS.

Kühne, O. (2019b). Vom ‚Bösen' und ‚Guten' in der Landschaft – das Problem moralischer Kommunikation im Umgang mit Landschaft und ihren Konflikten. In K. Berr & C. Jenal (Eds.), *Landschaftskonflikte* (pp. 131–142). Springer VS.

Kühne, O. (2020). Landscape Conflicts: A Theoretical Approach Based on the Three Worlds Theory of Karl Popper and the Conflict Theory of Ralf Dahrendorf, Illustrated by the Example of the Energy System Transformation in Germany. *Sustainability: Science, Practice and Policy, 12*(17), Article 6772. https://doi.org/10.3390/su12176772

Kühne, O. (2024). *Redescribing Horizontal Geographies: A Neopragmatist Approach to Spatial Contingency, Complexity, and Relationships*. Springer International.

Kühne, O. (2025). *Failing Beautifully in Science*. Springer Nature Switzerland. https://doi.org/10.1007/978-3-032-05336-7

Kühne, O., & Berger, S. (2026). Vokabulare, Konflikte und Demütigungen – das Beispiel der Veränderung materieller Räume und ihrer sozialen Resonanzen. In O. Kühne, K. Berr, & C. Jenal (Eds.), *Neopragmatismus und Raum* (in diesem Band). Springer VS.

Kühne, O., & Berr, K. (2021). *Wissenschaft, Raum, Gesellschaft: Eine Einführung zur sozialen Erzeugung von Wissen*. Springer VS.

Kühne, O., & Berr, K. (2023). Tourismus, Landschaft und Konflikt – ein theoretischer Rahmen. In O. Kühne, T. Freytag, T. Sedelmeier, & C. Jenal (Eds.), *Landschaft und Tourismus* (pp. 101–117). Springer.

Kühne, O., & Berr, K. (2026). Der Neopragmatismus als raumwissenschaftliche Meta-Theorie. In O. Kühne, K. Berr, & C. Jenal (Eds.), *Neopragmatismus und Raum* (in diesem Band). Springer VS.

Kühne, O., Berr, K., & Jenal, C. (2022). *Die geschlossene Gesellschaft und ihre Ligaturen: Eine Kritik am Beispiel 'Landschaft'*. Springer VS.

Kühne, O., Berr, K., Lohmann, P., & Schuster, K. (2025). *Panrelationen und Vokabulare auf dem Weg zu einer neopragmatistischen Synthese für Sozial- und Raumwissenschaften*. Springer VS.

Kühne, O., Edler, D., Berr, K., Lohmann, P., & Schuster, K. (2026a). Die kausale Abhängigkeit von Welt und ihre kartographische Aufbereitung – eine neopragmatistische Zuwendung jenseits repräsentationaler Vorstellungen. In O. Kühne, K. Berr, & C. Jenal (Eds.), *Neopragmatismus und Raum* (in diesem Band). Springer VS.

Kühne, O., Müller, M., & Berr, K. (2026b). Lebenschancen zwischen privater Selbsterschaffung und öffentlicher Solidarität – ein Dialog zwischen Richard Rorty und Ralf Dahrendorf. In O. Kühne, K. Berr, & C. Jenal (Eds.), *Neopragmatismus und Raum* (in diesem Band). Springer VS.

Laclau, E., & Mouffe, C. (1985). *Hegemony and Socialist Strategy: Towards a Radical Democratic Politics*. Verso.

Lakoff, G., & Johnson, M. (2003 [1980]). *Metaphors We Live By: With a New Afterword*. University of Chicago Press.

Leerhoff, H., Rehkämpfer, K., & Wachtenhofer, T. (2010). *Einführung in die Analytische Philosophie*. WBG.

Lohmann, P., Kühne, O., & Berr, K. (2024). *Landschaft und Gefühl – eine neopragmatistische Redeskription auf Grundlage der Philosophie Fichtes: Ein Beitrag zur humanistischen Geographie*. SpringerVS.
Luft, S. (2019). Den Verstand bewahren: Warum der Moralismus die Gesellschaft in den Bürgerkrieg treibt. In S. Kostner (Ed.), *Identitätslinke Läuterungsagenda: Eine Debatte zu ihren Folgen für Migrationsgesellschaften* (pp. 209–220). Ibidem-Verlag.
Lübbe, H. (2019). *Politischer Moralismus: Der Triumph der Gesinnung über die Urteilskraft*. LIT.
Marquard, O. (2007). Freiheit und Pluralität. In *Skepsis in der Moderne: Philosophische Studien* (pp. 109–123). Reclam.
Merker, B., Mohr, G., & Siep, L. (Eds.). (1998). *Angemessenheit: Zur Rehabilitierung einer philosophischen Metapher*. Königshausen & Neumann.
Mouffe, C. (2007). Pluralismus, Dissens und demokratische Staatsbürgerschaft. In M. Nonhoff (Ed.), *Diskurs – radikale Demokratie – Hegemonie: Zum politischen Denken von Ernesto Laclau und Chantal Mouffe* (pp. 41–53). transcript.
Mouffe, C. (2014). *Agonistik: Die Welt politisch denken* (Vol. 2677). Suhrkamp.
Müller, M. (2014). *Private Romantik, öffentlicher Pragmatismus? Richard Rortys transformative Neubeschreibung des Liberalismus* (1st ed.). *Edition Moderne Postmoderne*. transcript. https://doi.org/10.5555/9783839420416
Müller, M. (2023). Einführung in das Handbuch Richard Rorty. In M. Müller (Ed.), *Handbuch Richard Rorty* (pp. 9–41). Springer Fachmedien.
Neuhäuser, C., & Seidel, C. (2020). Kritik des Moralismus: Eine Landkarte zur Einleitung. In C. Neuhäuser & C. Seidel (Eds.), *Kritik des Moralismus* (pp. 9–34). Suhrkamp.
Niedenzu, H.-J. (2001). Konflikttheorie: Ralf Dahrendorf. In J. Morel, E. Bauer, T. Maleghy, H.-J. Niedenzu, M. Preglau, & H. Staubmann (Eds.), *Soziologische Theorie: Abriß ihrer Hauptvertreter* (7th ed., pp. 171–189). R. Oldenbourg Verlag.
Ott, K. (Ed.). (1996). *Vom Begründen zum Handeln: Aufsätze zur angewandten Ethik*. Attempto-Verlag.
Ottmann, H. (2012). *Geschichte des politischen Denkens: Band 4: Das 20. Jahrhundert*. Teilband 2: Von der Kritischen Theorie bis zur Globalisierung. J.B. Metzler.
Popper, K. R. (1935). *Die Logik der Forschung*. Mohr Siebeck.
Popper, K. R. (1963). *Conjectures and Refutations: The Growth of Scientific Knowledge*. Routledge & Kegan.
Popper, K. R. (1965). *Das Elend des Historizismus* (L. Walentik, Trans.). *Die Einheit der Gesellschaftswissenschaften: Vol. 3*. Mohr Siebeck.
Popper, K. R. (1992). *Die offene Gesellschaft und ihre Feinde: Der Zauber Platons* (7th ed., Vol. 1). J.C.B. Mohr. (Original work published 1945)
Prange, K. (2010). *Die Ethik der Pädagogik: Zur Normativität erzieherischen Handelns*. Ferdinand Schöningh.
Prasad, M. (2021). Pragmatism as Problem Solving. *Socius: Sociological Research for a Dynamic World*, *7*, Article 2378023121993991. https://doi.org/10.1177/2378023121993991
Pufé, I. (2017). *Nachhaltigkeit* (3., überarbeitete und erweiterte Auflage). *UTB: Vol. 8705*. UVK.
Putnam, H. (1990). *Die Bedeutung von "Bedeutung"* (2., durchgesehene Auflage). Klostermann.
Putnam, H. (1995a). *Pragmatism: An Open Question*. Blackwell.
Putnam, H. (1995b). Why There Isn't a Ready-made World. In P. K. Moser & J. D. Trout (Eds.), *Contemporary Materialism: A Reader* (pp. 233–253). Routledge.
Putnam, H. (1997). *Für eine Erneuerung der Philosophie*. *Universal-Bibliothek: Vol. 9660*. Reclam.
Putnam, H. (2002). *The Collapse of the Fact/Value Dichotomy and Other Essays*. Harvard University Press.
Rawls, J. (1975 [engl. Original 1971]). *Eine Theorie der Gerechtigkeit*. Suhrkamp.
Rawls, J. (1993). *Political Liberalism*. Columbia University Press.
Rawls, J. (2017 [engl. Original 1971]). *Eine Theorie der Gerechtigkeit* (20th ed.). Suhrkamp.
Reese-Schäfer, W. (2023). Richard Rorty und die neue Selbstdeutung der Gerechtigkeitstheorie von John Rawls. In M. Müller (Ed.), *Handbuch Richard Rorty* (pp. 413–430). Springer Fachmedien.

Reijnierse, W. G., Brugman, B., & Droog, E. (2025). The Differential Effects of Metaphor on Comprehensibility and Comprehension of Environmental Concepts. *Journal of Science Communication, 24*(4). https://doi.org/10.22323/150520250702095506
Reusswig, F. (2019). Heimat und politische Parteien. In M. Hülz, O. Kühne, & F. Weber (Eds.), *Heimat: Ein vielfältiges Konstrukt* (pp. 371–389). Springer VS.
Reusswig, F., Braun, F., Heger, I., Ludewig, T., Eichenauer, E., & Lass, W. (2016). Against the Wind: Local Opposition to the German Energiewende. *Utilities Policy, 41*, 214–227. https://doi.org/10.1016/j.jup.2016.02.006
Rorty, R. (1979). *Philosophy and the Mirror of Nature*. Princeton University Press.
Rorty, R. (1981 [1979]). *Der Spiegel der Natur: Eine Kritik der Philosophie*. Suhrkamp.
Rorty, R. (1988). *Solidarität oder Objektivität? 3 philos. Essays. Reclams Universal-Bibliothek: Nr. 8513*. Reclam.
Rorty, R. (1989). *Contingency, Irony, and Solidarity*. Cambridge University Press.
Rorty, R. (1991a). *Essays on Heidegger and Others. Philosophical Papers: Vol. 2*. Cambridge University Press. https://doi.org/10.1017/CBO9780511609039
Rorty, R. (1991b). *Objectivity, Relativism, and Truth*. Cambridge University Press.
Rorty, R. (1997 [1989]). *Contingency, Irony, and Solidarity*. Cambridge University Press. https://doi.org/10.1017/CBO9780511804397
Rorty, R. (1998). *Truth and Progress: Philosophical Papers*. Cambridge University Press.
Rorty, R. (2000). *Philosophie & die Zukunft: Essays* (2nd ed.). Fischer-Taschenbuch-Verlag.
Rorty, R. (2001a). Erwiderung auf Friederike Müller-Frimauth. In T. Tietz, U. Schäfer, & R. Zill (Eds.), *Hinter den Spiegeln: Beiträge zur Philosophie Richard Rortys mit Erwiderungen von Richard Rorty* (pp. 259–263). Suhrkamp.
Rorty, R. (2001b). *Die Schönheit, die Erhabenheit und die Gemeinschaft der Philosophen. Erbschaft unserer Zeit: Vol. 5*. Suhrkamp.
Rorty, R. (2021 [1989]). *Kontingenz, Ironie und Solidarität* (13th ed.). Suhrkamp.
Rorty, R. (2023a). *Pragmatismus als Antiautoritarismus* (J. Schulte, Trans.) (E. Mendieta, Ed.). Suhrkamp.
Rorty, R. (2023b). *Pragmatismus als Antiautoritarismus* (E. Mendieta, Ed.). Suhrkamp.
Rosa, H. (2016). *Resonanz: Eine Soziologie der Weltbeziehung. Suhrkamp-Taschenbuch Wissenschaft: Vol. 2272*. Suhrkamp.
Schlottmann, A., & Rosa, H. (2023). Resonanzpädagogik. In E. Nöthen & V. Schreiber (Eds.), *Transformative Geographische Bildung: Schlüsselprobleme, Theoriezugänge, Forschungsweisen, Vermittlungspraktiken* (pp. 163–169). Springer. https://doi.org/10.1007/978-3-662-66482-7_24
Schmid, S., & Stoichita, P. (2010). 4. Baruch de Spinoza. In D. Perler & J. Haag (Eds.), *Ideen: Repräsentationalismus in der frühen Neuzeit. Band 1: Texte* (pp. 123–159). de Gruyter.
Schmitt, C. (1933). *Der Begriff des Politischen*. Hanseatische Verlagsanstalt.
Schmitt, C. (2011 [1967]). *Die Tyrannei der Werte* (3., korrigierte Auflage). Duncker & Humblot.
Shklar, J. N. (1985). *Ordinary Vices*. Belknap Press.
Shklar, J. N. (1990). *The Faces of Injustice: Storrs Lectures on Jurisprudence/Yale Law School: Vol. 1988*. Yale University Press.
Shklar, J. N. (2020). Der Liberalismus der Furcht. In H. Bajohr (Ed.), *Judith N. Shklar: Der Liberalismus der Furcht* (3rd ed., pp. 26–66). Matthes & Seitz.
Shusterman, R. (2022). Pragmatist Philosophy for Our Times: Reviewing Rorty's Legacy. *Society, 59*(5), 583–590. https://doi.org/10.1007/s12115-022-00769-x
Stederoth, D. (2011). Kritik. In P. Kolmer & A. G. Wildfeuer (Eds.), *Neues Handbuch philosophischer Grundbegriffe* (Vol. 2, pp. 1346–1357). Alber.
Tetens, H. (1999). Wissenschaft. In H. J. Sandkühler (Ed.), *Enzyklopädie Philosophie* (pp. 1763–1773). Meiner.
Thiel, A., & Kühne, O. (2024). *Theorien sozialer Konflikte – vom Überblick zur Weiterentwicklung*. Springer VS.
Thiel, A., & Kühne, O. (2025). *Theories of Social Conflict: From Overview to Conceptual Development*. Springer International.

Topper, K. (1995). Richard Rorty, Liberalism and the Politics of Redescription. *American Political Science Review, 89*(4), 954–965. https://doi.org/10.2307/2082520

Weber, F., Kühne, O., Jenal, C., Aschenbrand, E., & Artuković, A. (2018). *Sand im Getriebe: Aushandlungsprozesse um die Gewinnung mineralischer Rohstoffe aus konflikttheoretischer Perspektive nach Ralf Dahrendorf*. Springer VS.

Chapter 10
Conclusion

If we take a neo-pragmatic and life chances-oriented perspective, sustainable development cannot be understood as the process of achieving normative end goals or fulfilling fixed models. In a neo-pragmatic and life chances-oriented understanding, it is rather understood as a process by which people continuously redefine their relationship to themselves and the world. It is therefore not to be understood as a state to be achieved, but as an open sociocultural practice that is determined by its processes, its changing concepts, and its forms of social sensitivity. In this book, we have attempted to show that sustainable development can be particularly successful where transformations are designed in such a way that they remain reasonable, resonant, and reusable for people. The neo-pragmatic approach makes it clear that vocabularies—and thus also understandings of sustainable development—are always contingent, historically situated, and revisable. Vocabularies (including those of sustainable development) do not derive their meaning from an (assumed) correspondence with an 'objective' order of the world or from 'alternative-free' instructions for action derived from 'facts'. A neo-pragmatic and life-chance-oriented understanding of sustainable development gains its meaning through its usefulness for orientation in complex social contexts and their references to the 'world out there', on which we are causally dependent but which we cannot representationally access (Rorty, 1979).

A neo-pragmatic and life-chance-oriented understanding of sustainable development emphasizes the constitutive importance of a reflexive redescription that intertwines old and new concepts. This type of redescription is in turn bound to the criteria of freedom from humiliation, connectivity, and recognition. In this sense, sustainable development can be understood as a process that is not secured by claims to truth, but by the quality of the conversation it enables. A conversation that is inclusive.

From this perspective, the concept of life chances becomes a structuring foundation: options and (voluntarily entered into) ligatures form the coordinate system in which individual freedom and collective bonds relate to each other. Sustainable development requires both the expansion of individually available options and the stabilization of voluntary bonds. Ligatures understood in this way enable orientation,

O. Kühne et al., *Sustainable Development and Life Chances*, Neopragmatic Horizontal Geographies 3, https://doi.org/10.1007/978-3-032-20182-9_10

belonging, and the production of meaning without unreflectively excluding options, as is the case with ligatures that are imposed on people without giving them the opportunity to decide whether they want to follow them or not (see, for example, Kühne et al., 2025; Leonardi et al., 2025). In the understanding presented in this book, self-creation and solidarity are not opposites, but complementary conditions of transformation that enable people to locate themselves in change and at the same time help shape it. At this point, we also do justice to the central position of World 2, which it occupies in the transmission of World 3 and World 1.

Sustainable development, in turn, depends on structural foundations. In the sense of a neo-pragmatic and life-chance-oriented understanding, procedural and opportunity justice are of central importance here. However, they are only necessary, but not sufficient, conditions for a resonant relationship. Such a relationship also requires compassion:

- Fair procedures ensure the possibility of participation and avoid social humiliation.
- Securing and expanding opportunities makes it possible to participate meaningfully in these procedures.
- Compassion, as a cognitive and affective sensitivity to the perspectives of others, forms the social basis that holds both together.

Compassion forms the basis for the kind of resonance that is necessary to resolve conflicts, respect differences, and understand transformations not as moral pressure but as a joint search for solutions. The concepts we have drawn on or further developed in this book, namely compassion, resonance, options and ligatures, redescription, acceptibility, and recognition, do not constitute a closed theoretical framework, nor do we regard them as definitively outlined. We understand them as conceptual offerings that can be taken up, continued, and transformed in accordance with changing challenges and contexts in future research, political practices, and social negotiations. Sustainable development thus remains an open, constantly falsifiable process, supported by the ability to continually produce new descriptions on the one hand and, on the other, to translate old descriptions in such a way that they can be recycled to benefit new challenges.

This openness is one of the strengths of the neo-pragmatic understanding of sustainability. A neo-pragmatic and life-chance-oriented understanding of sustainable development does not propose definitive solutions, but rather highlights the conditions under which societies can remain capable of learning, dialogue, and fairness in their dealings. Sustainable development thus becomes a continuous practice of understanding. In this practice, freedom and connectedness, criticism and empathy, change and stability are related to each other and subjected to a constant process of readjustment, integration of new metaphors, and ultimately appreciative redescriptions.

References

Kühne, O., Berr, K., Edler, D., Lohmann, P., & Schuster, K. (2025). *Neopragmatism—Inverse Landscape—(Carto)graphic Representation: From Critique to Further Development.* Springer International.

Leonardi, L., Kühne, O., & Berr, K. (2025). Ligatures and Their Leib-Bound and Socially Mediated Leib-Bound Ties—Foundations for an Actualization of Ralf Dahrendorf's Concept of Life Chances with Recourse to Karl Popper's Theory of the Three Worlds and Its Extension. *Smp – Società Mutamento Politica, 16*(31), 143–153. https://doi.org/10.36253/smp-15432

Rorty, R. (1979). *Philosophy and the Mirror of Nature.* Princeton University Press.

MIX
Papier aus verantwortungsvollen Quellen
Paper from responsible sources
FSC® C105338

If you have any concerns about our products,
you can contact us on
ProductSafety@springernature.com

In case Publisher is established outside the EU,
the EU authorized representative is:
Springer Nature Customer Service Center GmbH
Europaplatz 3, 69115 Heidelberg, Germany

Printed by Libri Plureos GmbH
in Hamburg, Germany